Test Bank to accompany

Chemistry
& Chemical Reactivity
THIRD EDITION
Kotz & Treichel

Karen E. Eichstadt

Ohio University

Saunders Golden Sunburst Series
Saunders College Publishing
Harcourt Brace College Publishers

Fort Worth Philadelphia San Diego New York Orlando Austin
San Antonio Toronto Montreal London Sydney Tokyo

Eichstadt; Test Bank to accompany Chemistry and Chemical
Reactivity, 3E. Kotz & Purcell.

ISBN 0-03-001313-5

567 017 987654321

Preface

This TEST BANK includes approximately fifty questions in two formats, free response and multiple for each chapter. Answers to the questions are given at the end of the section.

The free response questions are written for different levels of difficulty. Some are appropriate for honors students or an out-of-class assignment. An emphasis has been placed on the interpretation of real laboratory data and the design of experiments in these exercises.

The multiple choice questions contain only one correct answer. The terms "all of these" and "none of these" are used infrequently. The questions may easily be reworded for other evaluation formats. The numbers and formulas may be changed for variety.

Students need access to a periodic table, solubility tables and other standard tables as in the appendix of CHEMISTRY & CHEMICAL REACTIVITY, Third Edition.

Thanks are due Dr. John C. Kotz for his stimulating conversations and philosophical leadership and Elizabeth Rosato of Saunders College Publishing for her cheerful guidance. Jennifer Stephan provided essential help in the compilation and finishing processes. In addition, I acknowledge the many students who have contributed through their performance to the data base of appropriate exam questions as well as my colleagues, Professors David Hendricker, Gary Pfeiffer, and Lauren McMills. I especially thank my husband, Frank, and our children, Andrew, Amy, and David, for their patience and support in the project, .

The manuscript has been prepared on a Macintosh® computer using Microsoft® Word 5.0, Microsoft ®Excel 4.0, SuperPaint™ 3.5 and ChemDraw™ 2.1.3. The document was printed on an Apple Laserwriter Pro630.

Karen E. Eichstadt
Department of Chemistry
Ohio University
Athens OH
July 1995

TABLE OF CONTENTS

Chapter 1
Matter and Measurement

Section A: Free Response.

1. Two students were asked to determine the density of an unknown liquid using the same equipment consisting of an analytical balance and a 25.0 mL graduated cylinder. Student A reported (correctly) the density of the liquid as 2.0 g/mL while Student B reported (correctly) the value as 2.00 g/mL. Explain how each student performed the determination to be able to report the density with 2 significant figures in one case and 3 significant figures in the other.

2. When dry and empty, a flask had a mass of 27.31 grams. When filled with distilled water at 25.0°C, it weighed 36.84 grams. After cleaning and drying, the flask was filled with chloroform and found to weigh 41.43 grams. At 25.0°C, the density of water is 0.9970 g/cm^3. What is the density of chloroform?

3. If equal masses of two elements are placed in the same container and there is no chemical change, is the result a compound or a mixture? Explain.

4. The density of a liquid was determined in the laboratory. The liquid was left in an open container overnight. The next morning the density was measured again and found to be greater than it was the day before. Is the liquid a pure substance or a mixture? Explain.

5. Calculate the sum of 2.000 and 98.00. Explain the rules for number expression which apply to your answer.

6. Is the following statement true? Explain fully.
   "Any sample made up of only one substance is homogeneous."

7. Are the masses 25. mg, 2.50 x 10^{-2} grams, and 25.00 x 10^3 µg the same? Explain.

8. The laboratory notebook of a student's determination of the density of a metal is reproduced below. Using the information, calculate the density of the metal. Comment on the design of the experiment.

Procedure: The sample was weighed. Then it was inserted into a weighed graduated cylinder containing some water. After the insertion of the metal sample, the volume of the water and metal was again determined.

Data:
Mass of solid metal :	21.525 g
Mass of empty container:	113.235 g
Volume of water before immersion:	40.45 mL
Volume of water after immersion:	42.50 mL

9. Consider the densities of the liquid substances listed. a) How will substance X, a solid, with density of 0.87 g/cm^3, behave when placed in each of the other pure liquids? b) Sketch and label a test tube containing Substances A, B, C, and X. Assume that the liquids do not mix or react with each other and that solid Substance X does not react with any of the liquids.

Liquid Substance	Density (g/cm^3)	Does Substance X float or sink?
A	0.75	
B	1.1	
C	0.96	

10. From the graphical analysis of mass and volume data for Element A and Element B, answer the following questions and explain your reasoning.
   a) Which element is more dense?
   b) Would Element B float or sink in the liquid xylene which has a density of 0.88 g/cm^3?
   c) Of the four elements listed with their densities, select one as Element A and one as Element B.
   Explain your choices.

   Na   0.971 g/cm^3
   Li   0.534 g/cm^3
   Al   2.70 g/cm^3
   Zn   7.12 g/cm^3

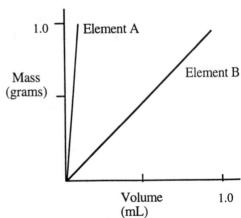

Answers to Free Response

The following answers contain the key concepts in the response. Students may have different, but correct, interpretations in some cases.

1. Student A used a smaller volume (<10.0 mL) while Student B used a larger volume (>10 mL). Thus each reported the correct number of significant figures using the same equipment.

2. Use the water data to find the volume of the flask which is the same in both cases.
$$(9.53 \text{ g})(1 \text{ cm}^3/0.9970\text{g}) = 9.56 \text{ cm}^3$$
   Find density of chloroform: $14.12\text{g}/9.56 \text{ cm}^3 = 1.48\text{g/cm}^3$

3. It must be a mixture because no new substance is formed.

4. The liquid was a mixture because the density changed overnight. If it had been a pure substance and some of it evaporated, the density would have still been the same.

5. The sum is 100.00. You are allowed two decimal places in the sum. In this case the fact that you now have 5 significant figures is not relevant to the question.

6. We need to know if the we have physical or chemical homogeneity. You could have a sample of pure H_2O at 0°C which is a mixture of ice and liquid water.

7. The masses are equivalent but have a different number of significant figures. This indicates that different tools were used to make these measurements.

8. The density is 10.5 g/mL. The student did not need to weigh the graduated cylinder because it is used only for the volume measurement. The small volume dictates only three significant figures in the reported density. Since the mass has 5 significant figures, using a larger sample size with a volume of more than 10 mL would improve the laboratory experiment.

9. Substance X will sink in A, float in B, and float in C.

10. Element A is more dense. Element B would sink in xylene because it has a mass/volume ratio greater than   0.88 g/cm³.

    Sodium most nearly matches the slope of the line which is nearly 1.
    Zinc has a steeper slope and approximates this line.

    Author's Note: This is a difficult question relating the density equation to a graph. In most cases students will not have seen this approach before. It offers students a chance to demonstrate their understanding on a new level.

Section B: Multiple Choice

11. Which physical state(s) of matter exhibit(s) the greatest change in volume with changes in temperature or pressure?

    a) none, they are all the same   b) liquids and gases change the same

    c) solid   d) liquid   e) gas

12. For most substances the volume of a mass of a gas is

    a) much larger than an equal mass of the liquid.

    b) slightly larger than an equal mass of the liquid.

    c) much smaller than an equal mass of a liquid.

    d) slightly smaller than an equal mass of a liquid.

    e) the same as an equal of a liquid.

13. The smallest particle of an element that retains the chemical properties of the element is a(n)

    a) atom   b) molecule   c) ion   d) solid   e) gas

14. According to the Kinetic Molecular Theory particles of a solid

    a) are bound in a regular array and do not move.

    b) float freely within an array occupying various positions relative to neighbors.

    c) have no relationship to the microscopic structure of the solid.

    d) vibrate back and forth but do not move past immediate neighbors.

    e) float freely in the inside but no not move on the surface.

15. In the gaseous state, particles

    a) move independently and randomly.

    b) have strong attractions for each other.

    c) fill the available space because of strong interactions.

    d) lose energy in time.

    e) gain energy in time.

16. When 10 grams of Liquid A is mixed with 40 grams of Liquid B the resulting solution has a density of 1.4 g/mL. What conclusion is true?

a) The density of Liquid A is greater than the density of Liquid B

b) The density of Liquid B is greater than the density of Liquid A.

c) Both liquids are more dense than water.

d) At least one of the liquids is more dense than water.

e) Neither liquid is more dense than water.

17. What is the density of a metal if a 15.4 gram sample has a volume of 1.96 cm^3?

a) 0.511g/cm^3 b) 0.127 g/cm^3 c) 7.86 g/cm^3

d) 30.2 g/cm^3 e) 33.1 g/cm^3

18. What volume of a liquid having a density of 3.48 g/cm^3 is needed to supply 5.00 grams of the liquid?

a) 1.44 cm^3 b) 17.4 cm^3 c) 0.696 cm^3

d) 5.75 cm^3 e) 6.71 cm^3

19. The density of aluminum is 2.70 g/cm^3. If a cube of aluminum weighs 13.5 grams, what is the length of the edge of the cube?

a) 5.00 cm   b) 1.71 cm   c) 1.25 cm

d) 0.200 cm   e) 0.3.12 cm

20. The dimensions of a rectangular solid are 8.45 cm long, 4.33 cm wide and 2.85 cm high. If the density of the solid is 9.43 g/ cm^3, what is its mass?

a) 1.12 grams   b) 11.1 grams   c) 154 grams

d) 896 grams   e) 983 grams

21. A metal sample weighing 30.9232 grams was added to a graduated cylinder containing 23.20 mL of water. The volume of the water plus the sample was 24.80 mL. What is the density of the metal?

a) 19.3 g/mL   b) 19.33 g/mL   c) 19.327 g/mL

d) 2.0 x 10^2 g/mL   e) 19. g/mL

22. Which temperature change is the smallest?
   a) 10°C to 20°C   b) 10 K to 20°C   c) 10K to 20K
   d) 10°F to 20°C   e) 10°F to 20°F

23. In comparing the Kelvin and Celsius temperature scales which one of the following
    statements is true?
    a) The difference between the freezing and boiling points of water is 273°C.
    b) The Kelvin scale has no negative numbers.
    c) The scales have the same zero point
    d) −273K = 0°C.
    e) 0 K = 273°C.

24. Approximately how many elements are found in nature?
    a) 185   b) 110   c) 90   d) 60   e) 44

25. How many significant figures are in the expression 2.430×10^2 grams?
    a) 3   b) 4   c) 5   d) 6   e) 7

26. The value seven hundred fifty million with four significant figures is written as
    a) 7500×10^6 b) 7.500×10^8 c) 750×10^6
    d) 750,000,000   e) 7500×10^8

27. Evaluate the expression $(8.346 + 2.854)(1.2750)=$
    a) 14.28   b) 1.428×10^1 c) 1.428×10^2
    d) 1.428   e) 1.4280×10^1

28. What is the numerical value of 1.5 cm $- 7.222 \times 10^{-1}$ cm?
    a) 0.7778 cm   b) 0.778 cm   c) 0.78 cm
    d) 0.8 cm   e) 7.072×10^{-1} cm

29. When the masses 0.0222 kg, 140.000 g and 5888 mg are added, the total should be
    reported with ___?___ significant figures.
    a) 3   b) 4   c) 5   d) 6   e) 7

30. Approximate the length of the line at the right.   _____

   a) 3 km   b) 3 m   c) 3 cm   d) 3 mm   e) 3 µm

31. Which one of the following elements is an alkali metal?

   a) Al   b) Ag   c) Au   d) Pb   e) Rb

32. Which of the following lists gives the symbol of a transition metal, a metal, and a non-metal in that order?

   a) Fe, Cl, Ca   b) Na, Fe, Cl   c) Fe, In, Ar

   d) Na, In, Cl   e) Ca, Fe, Ar

33. The symbols for a metal, a non-metal and a noble gas in that order are

   a) Ag, Ga, Xe   b) Ce, Ge, Ne   c) Ca, Sn, Kr

   d) Ba, P, Ar   e) P, Pb, Kr

34. Which of the following elements is a non-metal?

   a) Ca   b) Cr   c) Co   d) Cl   e) Cs

35. Which element is a gas at room temperature?

   a) argon   b) calcium   c) sulfur   d) lead   e) neodymium

36. A good example of an ionic compound is

   a) water   b) sugar   c) motor oil

   d) table salt   e) natural gas

37. Classify diamond, graphite, motor oil, and charcoal as element or compound.

   a) elements: diamond, charcoal   compounds: motor oil, graphite
   b) elements: graphite, diamond   compounds: charcoal, motor oil
   c) elements: motor oil, graphite   compounds: diamond, charcoal
   d) elements: diamond, charcoal, graphite compound: motor oil
   e) elements: motor oil, diamond   compounds: graphite, charcoal

38. When a pure solid substance was heated, a student obtained another solid and a gas, each of which was a pure substance. From this information which of the following statements is ALWAYS a correct conclusion?
    a) The original solid is not an element.
    b) Both products are elements.
    c) The original solid is a compound and the gas is an element.
    d) The original solid is an element and the gas is a compound.
    e) Both products are compounds.

39. Which of the following is a chemical property of water?
    a) its density is 1.000 g/cm^3 at 4°C.
    b) its melting point is 0°C
    c) it forms bubbles when calcium is added
    d) it causes light rays to bend
    e) its heat of fusion is 6020 J/mol

40. Which of the following is a *physical* combination of two or more pure substances?
    a) salt   b) air   c) sand   d) water   e) natural gas

41. To what category would a sample containing two substances and one phase belong?
    a) compound   b) element   c) diatomic mixture
    d) homogeneous mixture   e) heterogeneous mixture

42. Which of the following is an example of a chemical change?
    a) water boiling   b) ice melting   c) natural gas burning
    d) iodine vaporizing   e) dry ice subliming

43. Classify each observation as a physical or a chemical property and tally them.

        Observation 1: Bubbles form on a piece of metal when it is dropped into acid.

        Observation 2: The color of a crystalline substance is yellow.

        Observation 3: A shiny metal melts at 650ºC.

        Observation 4: The density of a solution is 1.84 g/cm^3

    a) 2 chemical properties and 2 physical properties

    b) 3 chemical properties and 1 physical properties

    c) 1 chemical properties and 3 physical properties

    d) 4 chemical properties

    d) 4 physical properties

44. To convert a value in kilograms to centigrams one should

    a) multiply by 10^5 b) multiply by 10^3 c) multiply by 10^{-3}

    d) divide by 10^5 e) divide by 10^{-1}

45. How many inches are in a length of 404 mm?

    a) 15.9 in   b) 103 in   c) 1.026 x 10^2 in

    d) 159 in   e) 1026 in

46. How many cm^2 are in an area of 4.21 in^2?

    a) 10.7 cm^2 b) 114 cm^2 c) 27.2 cm^2

    d) 1.66 cm^2 e) 1.14 cm^2

47. A wavelength of 17 nanometers is equal to ___?___ kilometers.

    a) 17 x 10^9 km   b) 17 x 10^{-9} km   c) 17 x 10 ^{12} km

    d) 17 x 10^{-12} km   e) 1.7 x 10^{-10} km

48. A pressure of 1.00 lbs/in^2 is equal to ___?___ .

    a) (454 g)^2 ÷ 2.54 cm   b) 454 g ÷ (2.54 cm)^2

    c) (454 g ÷ 2.54 cm)^2 d) 2.54 cm^2 ÷ 454 g

    e) [454 g ÷ (2.54 cm)^2]^2

49. When the prefix *micro* (μ) is used in the metric system, a fundamental unit of measurement is multiplied by a factor of

    a) 10^{-3} b) 10^6 c) 10^9 d) 10^3 e) 10^{-9}

50. A rectangular piece of aluminum foil is measured as 2.00 cm long and 3.45 cm wide. What is the area of the sheet?

    a) $7\,cm^2$ b) $6.9\,cm^2$ c) $6.90\,cm^2$

    d) $6.9000\,cm^2$ e) $7.0\,x\,10^2\,cm^2$

51. Of the masses 86.30 grams, 0.0863 kg and $8.630\,x\,10^5$ mg , which (if any) is the largest?

    a) 86.30 grams   b) 0.0863 kg   c) $8.630\,x\,10^5$ mg

    d) they are the same   e) two are the same, one is smaller

52. Consider the two statements regarding compounds and mixtures of elements:

    Statement 1: A compound has different properties than the elements of which it is
              composed.
    Statement 2: A compound has a definite percentage composition by mass of its
              combining elements.
    Statement 3: A compound must be composed of molecules.

    a) Only Statement 1 is true.   b) Only Statement 2 is true.

    c) Only Statement 3 is true   d) Statements 1 and 2 are true

    e) All three Statements are true.

53. A mixture of 80% nitrogen and 20% oxygen which can be separated into these elements by cooling is commonly called

    a) air   b) smog   c) an amino acid

    d) an acid   e) laughing gas

54. The element chlorine is obtained for commercial use in which of the following
    manners?
    a) Isolation from gas pockets in the earth's crust.
    b) Separation from air by a high pressure technique.
    c) Filtration of brine (NaCl) solutions.
    d) Electrolysis of aqueous NaCl solutions.
    e) Mixing sulfur and argon in equal quantities.

55. Brass is an alloy containing 66% copper and 34% zinc. How many grams of zinc are
    present in a 125 kg sample of the alloy?
    a) 2.4 grams   b) 43 grams   c) 83 grams
    d) 4.3 x 10^4 grams   e) 2.4 x 10^4 grams

56. If a 4.50 gram sample of a Zn-Al-Cu alloy contains 2.45 g Zn and 1.34 grams Al,
    what is the % composition of Cu?
    a) 3.8 %   b) 7.1%   c) 15.7%
    d) 54.4%   e) 71.0%

57. Tin foil is 92.0% tin and 8.00% zinc. How many kilograms of tin are in a 3.25 lb roll
    of tin foil? (1 kg = 2.20 lbs)
    a) 1.36 kg   b) 1.48 kg   c) 6.78 kg
    d) 7.15 kg   e) 11.8 kg

58. The density of a sodium sulfate solution is 1.32 g/cm^3. The solution is 8.22%
    sodium sulfate by mass. How many cm^3 of the solution are needed to supply 5.44
    grams of sodium sulfate?
    a) 10.8 cm^3 b) 4.12 cm^3 c) 33.8 cm^3
    d) 50.1 cm^3 e) 64.3 cm^3

59. Which of the following is NOT an SI base unit?
    a) mass   b) volume   c) length
    d) time   e) temperature

Chapter 1: Answers to Multiple Choice:

11. e	21. b	31. e
12. a	22. e	32. c
13. a	23. b	33. d
14. d	24. c	34. d
15. a	25. b	35. a
16. d	26. b	36. d
17. c	27. e	37. d
18. a	28. d	38. a
19. b	29. b	39. c
20. e	30. c	40. b

41. d	51. d
42. c	52. e
43. c	53. a
44. a	54. c
45. a	55. b
46. c	56. c
47. d	57. a
48. b	58. d
49. e	59. b
50. c	

Chapter 2
Atoms and Elements

Section A: Free Response

1. An element obtained from natural sources was analyzed by mass spectrometry. The graph at the right was obtained.

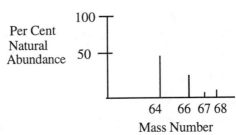

a) Identify the element. Explain how you reached this conclusion.

b) Using the experimental results of the mass spectrometry experiment, discuss information that is gained from the experiment that is not known from the information in the periodic table.

2. The elements hydrogen and oxygen have the naturally occurring isotopes in the abundances shown in the table. Based on this information, draw the anticipated mass spectrum for the gaseous molecule H_2O. Label appropriately.

Element	Mass Number	% Natural Abundance
hydrogen	1	99.98
hydrogen	2	0.02
oxygen	16	99.76
oxygen	18	0.20

3. As illustrated in the table below, the atomic number of an element is always an integer (whole number) but the atomic weight is seldom an integer (has decimal places). Explain.

Element	Atomic Number	Atomic Weight
Boron	5	10.81
Calcium	20	40.08
Tin	50	118.69

4. Suppose that fifty years from now scientists discover that Avogadro's number is actually somewhat larger than the value we now use. Would the statement below still be used by chemists appropriately? Explain fully.

     "One mole of carbon atoms weighs 12.01 grams."

5. Chlorine has two naturally occurring isotopes, Cl-35 and Cl-37. Using data from the Periodic Table determine which isotope is more abundant. Discuss their differences and similarities

6. About 200 years ago John Dalton published a theory on the composition of matter which has served as a foundation for modern chemistry. Is his postulate below still valid? If so, explain. If not, explain how we now have revised the statement. "Atoms of a given element are identical in mass and in properties."

7. Rank the following in order of increasing number of atoms.
     10.0 grams Zn   5.00 grams Ar   30.0 gram Pb   8.00 grams Fe

8. How many atoms are in an alloy prepared by melting 10.0 grams copper with 5.00 grams zinc and 2.00 grams lead, mixing thoroughly, and allowing the liquid to solidify?

9. Explain how the discovery of radioactivity and subsequent experiments in the years 1895-1900 required an extension of Dalton's atomic theory put forth in the year 1803.

10. The concept of periodicity was proposed by Mendeleev about the year 1869 with his Table of the Elements. Discuss the unique features of his table that have made it useful over the years with numerous new discoveries.

Answers to Free Response

1. The element is zinc because the average of the naturally occurring isotopes would be the atomic weight, 65.38. No other element has an average in this range. The mass spectrometry experiment provides information regarding the number of isotopes (4) and their masses (64, 66, 67, 68) as well as the % abundance. This information is not provided on the periodic table.

2.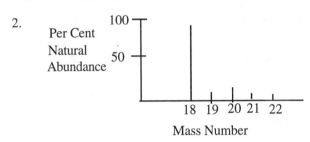

   Figure should show masses at 18 (largest), 19, 20 (2nd largest) , 21, and 22 for all combinations of isotopes of H and O in water.

3. The atomic number represents the number of protons in the nucleus and will be an integer. The atomic weight represents the average of naturally occurring isotopes and will not be a whole number. The varying number of neutrons in different abundances causes the average to be fractional.

4. Yes, the statement would still be correct but the definition of the number of "things" in a mole would be changed. One mole would weigh 12.01 grams but would merely contain more than 6.022×10^{23} atoms.

5. Since the average of the naturally occurring isotopes is 35.45 the Cl-35 isotope is more abundant (approximately 75%) and the Cl-37 isotope is about 1/3 as much (25%). The Cl-35 isotope has 17 protons and 18 neutrons while the Cl-37 isotope has 17 protons but 20 neutrons. The extra neutrons do not change the reactivity of the chlorine atoms appreciably.

6. Today, we know that atoms of a given element are not identical in mass because the number of neutrons can vary with the isotope. The number of protons, though, must remain the same. The properties of isotopes of an element are very similar.

7. 5.00 gram Ar < 8.00 g Fe < 30.0 g Pb < 10.0 g Zn

8. 1.46×10^{23} atoms

9. The discovery of radioactivity indicated that the atom was indeed divisible with smaller particles being emitted. Dalton's theory stated that the atom was indivisible. Later experiments regarding the discovery of protons, neutrons, and electrons allowed us to study parts of the atom. However, the mass changes studied by Dalton are still upheld.

10. Mendeleev arranged the elements known at the time into vertical columns based on their reactivity with oxygen, their density and other properties. In cases where he did not see similarity of properties, he left an empty space in the table assuming that at some point a new element with the predicted properties would be discovered. In every case, this was true. In addition the entire family of noble gases were not known to him. When they were discovered, they fit in at the end of each row.

Section B: Multiple Choice

11. In 1803, John Dalton's theory of postulated all of the following EXCEPT
    a) All matter is composed of atoms.
    b) All atoms of a given element are identical, but atoms of each element are different.
    c) Compounds are formed by the combination of two or more kinds of elements in small whole number ratios.
    d) Chemical reactions are the result of electrons being transferred from one atom to another.
    e) Atoms are not created, destroyed, divided into parts, or converted into other kids of atoms during a chemical reaction.

12. Which of the following sets of compounds illustrate the law of multiple proportions as set forth by John Dalton?
    a) CO_2, H_2O, O_2, H_2
    b) NO, NO_2, N_2O, N_2O_5
    c) CO_2, SO_2, SiO_2, TiO_2
    d) NO_2, NH_3, H_2NOH, H_2O
    e) C_2H_6, H_6C_2, CH_3CH_3, $(CH_3)_2$

13. If 4.00 grams of element A and 8.00 grams of element B react totally in a laboratory procedure which of the following statements is true?
    a) The reactions requires 4 moles of A to react with 8 moles of B.
    b) The maximum mass of product that can be formed is 12.00 grams.
    c) The maximum amount of product that can be formed is 12 moles.
    d) The atomic weight of B is double the atomic weight of A.
    e) The density of B is greater than the density of A.

14. When a reaction of A and B is performed in a sealed vessel, no change in the mass is observed. However, when the reaction is performed on the laboratory bench, the mass after the reaction is greater than the mass before the reaction. Based on this information, which of the following statements is true?
    a) Matter was created in the laboratory.
    b) An isotope of the reactant has been created.
    c) A gaseous product of the reaction was not weighed in the laboratory experiment.
    d) A gaseous reactant was not weighed in the laboratory experiment.
    e) Human error caused the discrepancies.

15. About the year 1910 Rutheford and colleagues performed experiments by targeting a stream of alpha particles(+ charged) at a piece of gold foil and recording the deflection of the particles on a sensitive screen. Which of the following statement(s) were conclusion(s) from those experiments?

     I. most of the volume of the atom is empty space
    II. the nucleus of an atom is dense and positively charged
   III. electrons have negligible mass

   a) I only   b) II only   c) I and II only
   d) II and III only   e) I, II, and III

16. In the Millikan oil drop experiment, the charge on oil droplets was observed by their behavior between charged plates as diagrammed in the figure. The fundamental change on an electron was determined as -1.60×10^{-19} coulombs by observing that

a) the charge on all the droplets was a multiple of 1.60×10^{-19} coulombs.

b) the charge on all the droplets was -1.60×10^{-19} coulombs.

c) the charge on all the droplets was $+1.60 \times 10^{-19}$ coulombs.

d) the charge on all the droplets was 1.60×10^{19} coulombs.

e) the charge on all the droplets was -1.60×10^{-19} coulombs

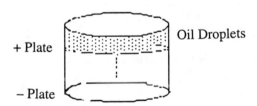

+ Plate

Oil Droplets

– Plate

17. The number of isotopes in a sample of a pure element can best be experimentally determined with a(n)

   a) electroscope   b) mass spectrometer   c) defraction grating
   d) cathode ray tube   e) scanning tunneling microscope

18. The masses 9.1094×10^{-28} grams, 1.6726×10^{-24} grams, and 1.6749×10^{-24} grams represent what particles respectively?

     a) proton, neutron, electron   b) electron, proton, neutron

     c) electron, neutron, proton   d) neutron, proton, electron

     d) proton, electron, neutron

19. Chlorine has two stable isotopes with exact masses of 34.96885 amu and 36.96590 amu. What is the relative abundance of the two isotopes?

     a) 50.00% ^{35}Cl and 50.00% ^{37}Cl b) 35.45% ^{35}Cl and 64.55% ^{37}Cl

     c) 64.55% ^{35}Cl and 35.45% ^{37}Cl d) 24.33% ^{35}Cl and 75.77% ^{37}Cl

     e) 75.77% ^{35}Cl and 24.33% ^{37}Cl

20. Boron has two stable isotopes with exact masses of 10.0129 amu and 11.0093 amu. What is the relative abundance of the two isotopes?

     a) 50.00% ^{10}B and 50.00% ^{11}B b) 19.91% ^{10}B and 80.09% ^{11}B

     c) 80.09% ^{10}B and 19.91% ^{11}B d) 10.81% ^{10}B and 89.19% ^{11}B

     e) 89.19% ^{10}B and 10.81% ^{11}B

21. Which of the following contains 24 neutrons?

     a) $^{52}_{24}Cr$ b) $^{24}_{12}Mg$ c) $^{24}_{48}Cd$ d) $^{48}_{24}Cr$ e) $^{24}_{24}Ca$

22. A nucleus which contains 25 protons and 25 neutrons is symbolized by

     a) $^{118.7}_{50}Sn$ b) $^{50}_{25}Sn$ c) $^{25}_{25}Mn$ d) $^{55.95}_{25}Mn$ e) $^{50}_{25}Mn$

23. How many neutrons are in $^{79}_{34}Se$?

     a) 113   b) 34   c) 45   d) 79   e) 11

24. Identify the set which contains isotopes of the same element.

     a) $^{17}_{8}A$ and $^{17}_{9}A$ b) $^{12}_{7}A$ and $^{7}_{12}A$ c) $^{11}_{7}A$ and $^{12}_{7}A$

     d) $^{11}_{5}A$ and $^{11}_{6}A^{+}$ e) $^{11}_{24}A$ and $^{24}_{11}A$

25. What is the mass number of an atom of iodine with 76 neutrons?

     a) 76   b) 53   c) 106   d) 129   e) 258

26. The names deuterium and tritium, respectively, refer to
    a) $_1^2\text{H}$ and $_1^3\text{H}$
    b) $_2^1\text{H}$ and $_3^1\text{H}$
    c) $_2^1\text{He}$ and $_3^1\text{He}$
    c) $_2^2\text{H}$ and $_2^3\text{He}$
    d) $_2^2\text{He}$ and $_2^3\text{He}$

27. The most abundant isotope of the element argon has ___?___ protons, ___?___ neutrons, and ___?___ electrons.
    a) 18, 40, 18
    b) 18, 22, 18
    c) 40, 18, 18
    d) 40, 39, 40
    e) 18, 20, 22

28. A transition metal, a halogen, and a metalloid in that order are
    a) Ni, N, Sn
    b) Sc, Si, Sb
    c) Cr, Cl, As
    d) As, Cl, Se
    e) Ca, Cl, Se

29. An alkaline earth, a chalcogen, and a noble gas in that order are
    a) Na, S, Ne
    b) Ca, O, Ne   c) K, Cl, Kr
    d) Sr, S, Se
    e) Mg, Br, Kr

30. Three elements in the lanthanide series are
    a) Ce, U, Rn
    b) Ce, Nd, Sm
    c) Ce, Ta, Nb
    d) Cs, Ba, Ce
    e) Fr, Ra, Ce

31. The chemical properties of sulfur would be most similar to
    a) P
    b) Cl
    c) Ar
    d) Se
    e) Ge

32. Which of the following elements is a non-metal?
    a) Ca
    b) Cr
    c) Co
    d) Cu
    e) Cl

33. A sample of 1.00 grams of sodium contains ___?___ moles.
    a) 2.16×10^{23}
    b) 2.62×10^{22}
    c) 1.25×10^{25}
    d) 22.99
    e) 0.0434

34. A sample of 20.0 grams of calcium contains ____?____ moles.
    a) 0.500   b) 1.000   c) 2.00
    d) 3.01×10^{23} e) 1.20×10^{24}

35. To obtain 1.50 moles of iron you must weigh ____?____ grams.
    a) 5.04×10^{25} b) 2.24×10^{25} c) 4.46×10^{-26}
    d) 37.2   e) 83.8

36. A sample of 1.00 grams of lead contains ____?____ atoms.
    a) 4.83×10^{-3} b) 1.25×10^{26} c) 6.02×10^{23}
    d) 207.2   e) 2.91×10^{21}

37. To obtain 1.20×10^{24} atoms of nickel, you would weigh
    a) 2.00 grams   b) 29.4 grams   c) 117 grams
    d) 7.04×10^{25} grams e) 1.42×10^{-26} grams

38. To obtain 5.66×10^{21} atoms of nickel, you would weigh
    a) 0.552 grams   b) 1.81 grams   c) 106 grams
    d) 9.64×10^{19} grams e) 1.04×10^{-20} grams

39. What is an expression for calculating the average mass of *one* atom of argon?
    a) 39.9 g / atom
    b) 1 g / 6.02×10^{23} atoms
    c) 39.9 g / 6.02×10^{23} atoms
    d) 6.02×10^{23} atoms / 39.9 g
    e) (39.9 g / 1 mol) (6.02×10^{23} atoms / 1 mol)

40. What is an expression for calculating the average mass of *one* atom of calcium?
    a) (40.08 g / 1 mol)(6.022×10^{23} atoms / 1 mol)
    b) (1 mol / 40.08 g)(6.022×10^{23} atoms / 1 mol)
    c) (40.08g / 1 mol)(1 mol / 6.022×10^{23} atoms)
    d) (1 mol / 40.08 g)(1 mol / 6.022×10^{23} atoms)
    e) (40.08 g / 1 atom)(6.022×10^{23} atoms / 1 gram)

41. What is the average mass of one atom of copper?
    a) 9.48×10^{-21} g   b) 1.05×10^{-22} g   c) 6.02×10^{-23} g
    d) 3.82×10^{25} g   e) 1.66×10^{-24} g

42. What is the average mass of one atom of neon?
    a) 3.35×10^{-23} g   b) 1.21×10^{-25} g   c) 6.02×10^{-23} g
    c) 2.98×10^{22} g   d) 1.66×10^{-24} g

43. If 2.00 µL of mercury which has a density of 13.53 g/mL is used in an experiment, how many moles and how many atoms are used?
    a) 1.34×10^{-2} mol, 8.06×10^{21} atoms
    b) 2.71×10^{-2} mol, 1.63×10^{22} atoms
    c) 6.76×10^{-3} mol, $4.07\ 10^{21}$ atoms
    d) 1.35×10^{-4} mol, 8.12×10^{19} atoms
    e) 6.74×10^{-5} mol, 1.12×10^{21} atoms

44. If 6.00 µL of argon which has a density of 1.78 g/L is used in an experiment, how many moles and how many atoms are used?
    a) 2.67×10^{-7} mol, 1.61×10^{17} molecules
    b) 1.34×10^{-2} mol, 6.21×10^{18} molecules
    c) 1.07×10^{-5} mol, 6.43×10^{18} molecules
    d) 1.18×10^{-11} mol, $1.97\ 10^{17}$ molecules
    e) 4.45×10^{-8} mol, 2.68×10^{16} molecules

45. The density of copper is 8.96 g/cm^3 at room temperature. If a cube of copper weigh 25.0 grams, what is the length of its edge?
    a) 0.635 cm   b) 3.91 cm   c) 6.79 cm
    d) 1.51 cm   e) 2. 07 cm

46. The density of zinc is 7.14 g/cm^3 at room temperature. How many atoms are in a cube of pure zinc which has an edge of 3.41 cm ?
    a) 1.16×10^{26} atoms   b) 1.97×10^{23} atoms   c) 3.48×10^{22} atoms
    d) 6.58×10^{22} atoms   e) 1.78×10^{24} atoms

47. The density of lead is 11.3 g/cm^3 at room temperature. How many atoms are in a
    cube of pure lead which has an edge of 2.00 cm ?
    a) 4.11 x 10^{21} atoms   b) 3.11 x 10^{20} atoms   c) 1.38 x 10^{24} atoms
    d) 2.63 x 10^{23} atoms   e) 6.57 x 10^{22} atoms

48. The density of argon is 1.78 g/L. What volume of argon would contain 3.01 x 10^{25}
    atoms?
    a) 447 liters   b) 22.4 liters   c) 11.2 liters
    d) 560 liters   e) 13.7 liters

49. The density of helium is 0.178 g/L. What volume of helium would contain
    8.45 x 10^{24} atoms?
    a) 5.07 x 10^2 liters   b) 3.77 x 10^{23} liters   c) 6.31 x 10^2 liters
    d) 2.65 x 10^{24} liters   e) 3.14 x 10^2 liters

50. How many grams of magnesium contain the same number of atoms as 20.04 grams of
    calcium?
    a) 12.15 g   b) 20.04 g   c) 24.30 g
    d) 40.08 g   e) 48.60 g

51. How many grams of magnesium contain the same number of atoms as 1.00 grams of
    calcium?
    a) 0.411 g   b) 0.606 g   c) 1.65 g
    d) 1.08 x 10^{-24} g   e) 9.93 x 10^{-23} g

52. How many grams of iron contain the same number of atoms as 50.0 grams of
    aluminum?
    a) 1.24 x 10^{24}g   b) 1.08 x 10^{-24}g   c) 80.2g
    d) 104 g   e) 111 g

53. The positive charge in the nucleus of an element determines the
    a) atomic mass   b) mass number   c) atomic number
    d) number of neutrons   e) radioactivity

54. There are two stable isotopes of carbon. They differ with respect to
    a) atomic mass   b) number of protons   c) radioactivity
    d) atomic number   e) electron configuration

55. A transition metal, a halogen, and a metalloid in that order are
    a) Ni, N, Sn   b) Sc, Si, Sb   c) As, Cl, Se
    d) Bi, Br, C   e) Cr, Cl, As

56. An alkaline earth, a chalcogen, and a noble gas in that order are
    a) Na, S, Ne   b) Ca, O, Ne   c) K, Cl, Kr
    d) Sc, S, Ra   e) Mg, Br, Kr

57. A lanthanide, a chalcogen, and a transition metal in that order are
    a) Tb, Tl, Tc   b) Sm, S, Sc   c) Cm, Cl, Cs
    d) U, O, Os   e) Am, As, Au

58. Which of the following elements are in the same chemical family?
    a) Rn, Ba, Sr, Be   b) N, O, F, Ne   c) Li, Be, Na Mg
    d) Ge, As, Sb, Te   e) Si, Sn, C, Pb

59. The chemical properties of sulfur would be most similar to
    a) P   b) Cl   c) Ar   d) Se   e) As

60. The chemical properties of germanium would be most similar to
    a) P   b) Ga   c) Si   d) Sb   e) Ho

Chapter 2: Answers to Multiple Choice:

11. d	21. b	31. d
12. b	22. e	32. e
13. b	23. c	33. e
14. d	24. c	34. a
15. c	25. d	35. e
16. a	26. a	36. e
17. b	27. b	37. b
18. c	28. c	38. a
19. e	29. b	39. c
20. b	30. b	40. c

41. b	51. b
42. a	52. d
43. d	53. c
44. a	54. a
45. d	55. e
46. e	56. b
47. d	57. b
48. c	58. e
49. e	59. d
50. a	60. c

Chapter 3
Molecules and Compounds

Section A: Free Response

1. The element chlorine has two naturally occurring isotopes in the approximate abundances shown in the table. Based on this information, explain why the mass spectrum for the molecule Cl_2 would be anticipated to have the pattern illustrated in the region 70-75.

Element	Mass Number	% Natural Abundance
chlorine	35	75
chlorine	37	25

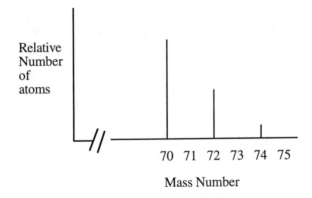

2. Consider the following substances and the amounts specified. Designate three which closely "match" and explain the basis for the similarity.

    18.02 g water
    80.60 g mercury
    44.01 g carbon dioxide
    70.90 g chlorine
    70.90 g bromine

3. Recognizing that O_2 and O_3 exist, with molar masses 32.0 and 48.0 respectively, would you expect scientists to discover another allotrope of oxygen with a molar mass of 40.0 g/mole? Explain why or why not.

4. The element arsenic (Atomic Number 33) is placed in a special category. The compounds As_2S_3, As_2S_5, and Ba_3As_2 are all listed in a chemical handbook. Name each compound and explain how these compounds illustrate the unique charter of the element arsenic.

5. Complete the following paragraph with appropriate numbers. *The length of the lines has no bearing on the proper answer!*

   A mole of nitrogen atoms contains _____ atoms. A mole of nitrogen molecules contains _____ molecules. A mole of nitrogen molecules contains _____ atoms. A mole of nitrogen atoms has a mass of _____ grams. A mole of nitrogen molecules has a mass of _____ grams.

6. The biopolymer DNA (deoxyribonucleic acid) is responsible for heredity. A sample of pure DNA has a density of 1.28 g/cm^3. The average molar mass of the DNA was $6.50 \times 10^8 g/mol$. What is the maximum volume, in cm^3, taken up by one DNA molecule?

7. What is the empirical formula of a compound which is 64.80% carbon, 6.35 % hydrogen, and 28.83 % sulfur?

8. A hydrate of lead (II) acetate is 14.2% water. Calculate the formula of the compound.

9. The weight percent of oxygen in an oxide that has the formula E_2O_3 is 30.06%. Calculate the atomic weight of element E and identify the probably element.

10. A *mixture* of KCl and K_2CO_3 was subjected to elemental analysis. The results showed that the mixture was 53.3% K, 38.0% Cl, 1.74% C and 6.93% O. Based on this information, calculate the mass percent KCl in the mixture.

Key Concepts for Free Response:

1. The largest peak at mass 70 represents a diatomic molecule with two Cl-35 atoms. This is the most likely possibility with 75% abundance. The next peak at mass 72 represents a molecule with one Cl-35 and one Cl-37. The smallest peak (and in least abundance) is the molecule formed from two Cl-37 atoms with a mass of 74.

2. 18.02 g water, 44.01 g carbon dioxide and 70.90 g chlorine are all 1 mole quantities and will have the same number of molecules.

3. No, any other combination of pure oxygen would have a greater mass, such as O_4 or O_5. A molar mass of 40 for a compound of pure oxygen would not be possible.

4. As_2S_3 is arsenic (III) sulfide with the arsenic as a metal having a 3+ charge.
    As_2S_5 is arsenic (V) sulfide with the arsenic as a metal having a 5+ charge.
    Ba_3As_2 is barium arsenide with arsenic as a non-metal having a 3- charge.
    These illustrate the unique character of arsenic as a metalloid with properties of both metals and non-metals.

5. A mole of nitrogen atoms contains 6.022×10^{23} atoms. A mole of nitrogen molecules contains 6.022×10^{23} molecules. A mole of nitrogen molecules contains 12.04×10^{23} atoms. A mole of nitrogen atoms has a mass of 14.01 grams. A mole of nitrogen molecules has a mass of 28.02 grams.

6. $8.43 \times 10^{-16} \text{ cm}^3$

7. C_6H_7S

8. $Pb(C_2H_3O_2)_2 \cdot 3H_2O$

9. The atomic weight is 55.9 so the element is iron and the oxide is Fe_2O_3.

10. The mixture is 79.8% KCl.

Section B: Multiple Choice

11. How many moles are in 5.00 grams of fluorine gas?

    a) 0.263 mol   b) 0.179 mol   c) 1.07×10^{23} mol

    d) 5.35×10^{23} mol   e) 3.80 mol

12. How many molecules are in 75.0 grams of nitrogen gas?

    a) 6.45×10^{24} molecules   b) 1.26×10^{27} molecules

    c) 8.06×10^{24} molecules   d) 1.61×10^{24} molecules

    e) 6.32×10^{26} molecules

13. How many atoms are in 40.0 grams of oxygen gas?

    a) 7.52×10^{23} atoms   b) 3.01×10^{24} atoms   c) 2.41×10^{25} atoms

    d) 1.25×10^{23} atoms   e) 1.51×10^{24} atoms

14. How many moles are in 8.50 mL of liquid bromine if the density of bromine is 3.12 g/mL?

    a) 0.166 mol   b) 0.0830 mol   c) 0.0723 mol

    d) 0.138 mol   e) 0.170 mol

15. The density of bromine is 3.420 g/mL. What is the volume of 0.486 moles?

    a) 161 mL   b) 551 mL   c) 47.1 mL

    d) 21.3 mL   e) 200. mL

16. A sample of oxygen gas weighs 32.0 grams. It contains <u>exactly</u> 6.022×10^{23} __?__ .

    a) protons   b) neutrons   c) electrons
    d) molecules   e) atoms

17. A mole of chlorine molecules weighs __?__ grams and contains __?__ atoms.

    a) 70.90 grams and 6.022×10^{23} atoms   b) 70.90 grams and 12.04×10^{23} atoms
    c) 35.45 grams and 6.022×10^{23} atoms   d) 35.45 grams and 12.04×10^{23} atoms
    e) 35.45 grams and 3.011×10^{23} atoms

18. How many moles are in 15.0 grams of potassium carbonate?

    a) 6.53×10^{22} mol   b) 2.07×10^{23} mol   c) 0.151 mol

    d) 0.109 mol   e) 9.20 mol

19. How many moles are in 1.07×10^{23} molecules of SO_2?

    a) 11.4 mol   b) 0.178 mol   c) 1.67×10^{21} mol

    d) 2.77×10^{-3} mol   e) 6.85×10^{24} mol

20. How many moles are in 1.07×10^{23} molecules of CO_2?

    a) 4.04×10^{-3} mol   b) 7.82 mol   c) 344 mol

    d) 5.62 mol   e) 0.178 mol

21. How many grams of CH_4 contain the same number of molecules as 2.50 grams O_2?

    a) 1.25 g   b) 0.0781 g   c) 0.156 g

    d) 4.70×10^{22} g   e) 4.88×10^{-3} g

22. What is the total number of *atoms* present in 25.0 mg of a compound $C_{10}H_{16}O$?

    a) 27 atoms   b) 1.51×10^{22} atoms   c) 9.89×10^{19} atoms

    d) 1.64×10^{-4} atoms   e) 2.67×10^{21} atoms

23. How many grams of NO_2 contain the same number of molecules as 5.00 grams H_2O?

    a) 12.8 g   b) 2.77 g   c) 5.00 g

    d) 0.391 g   e) 1.96 g

24. Which of the following has the same number of atoms as 55.8 grams of iron?

    a) 55.8 g H_2 b) 55.8 g He   c) 32.0 g O_2

    d) 16.0 g O_2 e) 32.0 g O_3

25. Which of the following compounds will form a solution with water that is a good
conductor of electricity?

    a) CCl_4 b) CO_2 c) NaCl

    d) Cl_2 e) CH_3OH

26. Which of the following compounds will form a solution with water that is a good conductor of electricity?

    a) CH_4 b) C_2H_6 c) $CaCl_2$

    d) SiH_4 e) CH_3OH

27. Some common ions based on metals are

    a) Fe^{2+}, Fe^{3+}, K^{2+} b) Mg^{2+}, Ba^{3+}, Na^+ c) Li^{2+}, Na^+, Al^{3+}

    d) Fe^{2+}, Sr^{2+}, Mg^{2+} e) Li^{2+}, Ca^{2+}, Al^{3+}

28. Some common ions based on non-metals are

    a) O^{2-}, Cl^{2-}, K^- b) O^{2-}, Cl^{2-}, N^{3-} c) S^{2-}, N^{2-}, Cl^-

    d) O_2^-, O^{2-}, P^{2-} e) S^{2-}, P^{3-}, F^-

29. Butane is a molecule with four carbon atoms and ten hydrogen atoms. The formula of butane is

    a) C_4H_{10} b) $(C_4H_5)_{10}$ c) C_2H_5

    d) $C_{10}H_4$ e) $(C_2H_{10})_5$

30. Glucose is a molecule with six carbon atoms, twelve hydrogen atoms and six oxygen atoms.   The formula of glucose is

    a) CH_2O b) $C_6H_{12}O_6$ c) $C(H_2O)_6$

    d) $(C_3H_{12}O_3)_6$ e) $C_3H_6O_3$

31. The calcium ion, Ca^{2+}, has

    a) two more electrons than the calcium atom

    b) two less electrons than the calcium atom

    c) two more protons than the calcium atom

    d) two less protons than the calcium atom

    e) two more electrons and two more protons than the calcium atom

32. The most common ION of atomic fluorine contains ___?___ electrons

        a) 8   b) 9   c) 10   d) 19   e) 20

33. The following species F^-, Ne, Na^+, and Mg^{2+} all have the same number of

    a) protons   b) neutrons   c) electrons

    d) nucleons   e) charges

34. For an atom from Group 7A of the Periodic Table, the most common monatomic ion
    of this atom will have a charge of
    a) +7   b) –7   c) +8   d) –1   e) -2

35. Which of the following pairs have the same number of electrons?
    a) Fe^{2+} and Fe^{3+} b) Ca^{2+} and K^+ c) K^+ and Na^+
    d) O_2^- and O^{2-} e) H^+ and H^-

36. Which of the following pairs have the same number of electrons?
    a) Cu^{2+} and Cu^+ b) Ca^{2+} and Mg^{2+} c) Cl^- and Br^-
    d) F^- and O^{2-} e) H^+ and H^-

37. Which formula represents the binary compound formed by magnesium and nitrogen?
    a) MgN   b) Mg_2N c) MgN_3
    d) Mg_3N_2 e) Mg_2N_3

38. Which formula represents the binary compound formed by sodium and sulfur?
    a) NaS   b) Na_2S c) NaS_3
    d) Na_3S_2 e) Na_3S

39. Which formula represents the binary compound formed by sodium and tellurium?
    a) Na_2Te b) NaTe   c) Na_2Te
    d) Na_3Te_2 e) Na_3Te

40. Which formula represents the binary compound formed by sodium and sulfur?
    a) NaS   b) Na_2S c) NaS_3
    d) Na_3S_2 e) Na_3S

41. Which group of compounds are ALL ionic?
    a) H_2O, NaCl, CS_2 b) NaCl, CH_4, $CaCl_2$
    c) $CaCl_2$, $FeCl_3$, NaCl   d) H_2O, $FeCl_3$, CO_2
    e) $CaCl_2$, $FeCl_3$, CO_2

42. Give the ions present and their relative numbers in potassium sulfate.
    a) 1 K^+ and 1 SO_4^- b) 2 K^+ and 1 SO_3^{2-} c) 1 K^+ and 2 SO_4^{2-}
    d) 2 K^+ and 1 SO_4^{2-} e) 3 K^+ and 1 SO_4^{3-}

43. Give the ions present and their relative numbers in barium nitrate.
    a) 1 Ba^+ and 1 NO_3^- b) 2 Ba^+ and 1 NO_3^{2-} c) 1 Ba^{2+} and 2 NO_3^-
    d) 1 Ba^{2+} and 1 NO_2^{2-} e) 1 Ba^{2+} and 2 NO_2^{2-}

44. What is the correct name for Cr_2O_3?
    a) chromium (VI) oxide   b) chromium (III) oxide   c) dichromium trioxide
    e) chromium (II) oxide   e) chromium (IV) oxide

45. What is the correct name for $FeSO_4$?
    a) iron (III) sulfite   b) iron (II) sulfite   c) iron (II) sulfate
    d) iron (II) sulfide   e) iron (III) sulfide

46. What is the correct name for Na_2S?
    a) sodium disulfide   b) disodium sulfide   c) sodium sulfide
    d) disodium sulfate   e) sodium sulfate

47. What is the correct name for $KClO_4$?
    a) potassium chlorate   b) potassium perchlorite
    c) potassium perchlorate   d) potassium hypochlorate
    e) potassium hypochlorite

48. What is the correct formula for sodium bicarbonate?
    a) $Na(CO_3)_2$ b) $NaHCO_3$ c) Na_2CO_3
    d) $NaCO_2$ e) $NaHCO_2$

49. The formula of the compound ammonium phosphate is
    a) $NH_4(PO_4)_3$ b) $NH_4(PO_4)_2$ c) $(NH_4)_2 PO_4$
    d) $(NH_4)_3PO_4$ e) $(NH_4)_3(PO_4)_2$

50. Which of the following compounds is osmium (VIII) oxide?
    a) OsO_4 b) Os_8O c) Os_4O_8
    d) Os_8O_2 e) Os_4O_2

51. What is the correct formula for sodium telluride?
    a) NaTe   b) $NaTeO_4$ c) Na_2Te
    d) Na_2TeO_4 e) Na_2Te_2

52. Which combination of name and formula is correct?
    a) aluminum phosphate, $Al_2(PO_4)_3$
    b) ammonium sulfate, NH_4SO_4
    c) iron (III) chloride, Fe_2Cl_3
    d) carbon tetrachloride, CCl_4
    e) magnesium nitrite, $Mg(NO_2)_2$

53. Which combination of name and formula is correct?
    a) diphosphorus pentoxide $P_2(O_2)_5$
    b) magnesium carbonate, $MgCO_3$
    c) iron (III) oxide, Fe_3O_2
    d) sodium sulfate, Na_2SO_4
    e) calcium chloride, Ca_2Cl

54. The formula of barium molybdate is $BaMoO_4$. Therefore, the formula of sodium molybdate is
    a) Na_4MoO b) $NaMoO$ c) Na_2MoO_3
    d) Na_2MoO_4 e) Na_4MoO_4

55. Sodium oxalate has the chemical formula $Na_2C_2O_4$ Based on this information, the formula of calcium oxalate is
    a) Ca_2CO_3 b) CaC_2O_4 c) $Ca_2C_2O_4$
    d) $CaHCO_3$ e) $CaCO_3$

56. Which of the following compounds is 36.4% nitrogen?
    a) N_2O b) NO c) N_2O_3
    d) N_2O_4 e) N_2O_5

57. The empirical formula for a compound is C_2H_3. Which of the following could be a molecular formula for the compound?
    a) CH_3 b) C_3H_4 c) $C_{12}H_{18}$
    d) $C_{30}H_{20}$ e) C_4H_9

58. The molar mass of barium nitrate is
    a) 199.21 g/mol   b) 229.32 g/mol   c) 261.32 g/mol
    d) 336.61 g/mol   e) 398.62 g/mol

59. The molar mass of sodium chlorate is
    a) 58.44 g/mol   b) 81.45 g/mol   c) 93.90 g/mol
    d) 106.45 g/mol   e) 129.45 g/mol

60. The molar mass of barium hydroxide octahydrate is
    a) 189.34 g/mol   b) 281.38 g/mol   c) 298.38 g/mol
    d) 299.48 g/mol   e) 315.48 g/mol

61. The molar mass of sodium carbonate decahydrate is
    a) 101.03 g/mol   b) 106.01 g/mol   c) 166.02 g/mol
    d) 286.21 g/mol   e) 315.48 g/mol

62. The molar mass of a compound with an empirical formula of BH_3 is 27.682 g/mol.
    What is the molecular formula?
    a) B_2H_6 b) B_2H_3 c) B_3H_6
    d) B_2H_4 e) B_2H_5

63. All of the compounds listed below have a molar mass of 240 amu. Which of the
    listed compounds could have an empirical mass of 80 amu?

    a) $C_{14}H_8O_4$ b) $C_{10}H_8OS_3$ c) $C_{15}H_{12}O_3$
    d) $C_{12}H_{18}O_4N$ e) $C_6H_8Br_2$

64. Calculate the empirical formula of glycine which is 32.00% carbon, 6.72% hydrogen,
    18.66% nitrogen and 42.63% oxygen.
    a) $C_2H_5NO_2$ b) $C_3H_6N_2O_4$ c) $C_2H_6N_2O_4$
    d) $C_3H_6N_2O$ e) $C_2H_6NO_2$

65. Calculate the empirical formula of a compound which is 51.40% carbon, 8.63% hydrogen, and 39.97% nitrogen.

   a) $C_5H_8N_2$ b) $C_4H_5N_2$ c) C_3H_5N
   d) $C_4H_4N_2$ e) $C_3H_4N_2$

66. Calculate the empirical formula a compound which is 39.12% carbon, 8.75% hydrogen, and 52.12% oxygen.

   a) $C_2H_5O_2$ b) $C_4H_8O_5$ c) $C_5H_9O_4$
   d) $C_3H_8O_3$ e) $C_4H_9O_2$

67. Calculate the empirical formula of a compound which is 50.69% carbon, 7.09% hydrogen, 19.71% nitrogen and 22.51% oxygen.

   a) $C_2H_5NO_2$ b) $C_3H_5N_2O_4$ c) C_3H_5NO
   d) C_2H_6NO e) $C_2H_6NO_2$

68. Calculate the empirical formula of a compound which is 76.34% carbon, 6.41% hydrogen, and 17.25% fluorine.

   a) $C_2H_5NO_2$ b) $C_3H_6N_2O_4$ c) $C_2H_6N_2O_4$
   d) $C_3H_6N_2O$ e) $C_2H_6NO_2$

69. A 3.26 g sample of an organic compound was found to contain 2.42 g carbon, 2.82 g hydrogen, and .563 g nitrogen. Calculate the empirical formula of the compound.

   a) $C_5H_4N_3$ b) $C_{10}H_2N_3$ c) $C_6H_7N_4$
   d) C_2H_3N e) C_5H_7N

70. Serotonin is a neurotransmitter which plays a role as neurons carry messages. The molar mass of serotonin is 176 g/mol. When a 5.31 g sample is analyzed, it was found to contain 3.62 g carbon, 0.362 g hydrogen, 0.844 g nitrogen and 0.482 g oxygen. What is the molecular formula of serotonin?

   a) $C_9H_{10}N_3O$ b) $C_{11}H_{14}NO$ c) $C_{10}H_{12}N_2O$
   d) $C_9H_8N_2O_2$ e) $C_{10}H_{26}NO$

Chapter 3: Answers to Multiple Choice:

11. b	21. a	31. b
12. d	22. e	32. c
13. e	23. a	33. c
14. a	24. d	34. d
15. c	25. c	35. b
16. d	26. c	36. d
17. b	27. d	37. d
18. d	28. e	38. b
19. b	29. a	39. a
20. e	30. b	40. b

41. c	51. c	61. d
42. d	52. d	62. a
43. c	53. d	63. c
44. b	54. c	64. a
45. c	55. b	65. e
46. c	56. a	66. d
47. c	57. c	67. c
48. b	58. c	68. a
49. d	59. d	69. e
50. a	60. e	70. c

Chapter 4
Principles of Reactivity: Chemical Reactions

Section A: Free Response

Students may have access to a Generalized Solubility Table

1. A water sample is thought to be contaminated with either silver ions or barium ions or both. Explain how you could use solutions of H_2SO_4 and $NaCl$ to determine conclusively what contaminant(s) are present. Assume you have an adequate supply of the contaminated sample to run as many tests as necessary.

2. When solutions of barium chloride and sulfuric acid are mixed, a white precipitate is formed and the aqueous layer is colorless. After being decanted the aqueous solution is a good conductor of electricity. Name the precipitate and explain why the decanted solution conducts an electric current.

3. You have been asked to identify four white solids in the laboratory. They are known to be barium chloride, sodium chloride, barium carbonate and sodium carbonate. The only reagents you have available are distilled water and a solution of hydrochloric acid. Design a procedure to identify the solids. Assume you have an adequate supply of the white solids and reagents.

4. Select reagents from those listed which could be used to prepare the indicated products with no contamination.   Explain how you would use the reagents and any additional procedure that may be necessary.

Available Reagents		
#1 AgCl (s)	#2 HNO_3 (aq)	#3 HCl (aq)
#4 $Sr(NO_3)_2$(aq)	#5 $SrCO_3$(s)	#6 Na_2SO_4(aq)
#7 NaOH (aq)	#8 $PbSO_4$(s)	

Product Desired
a) NaCl(aq)
b) $SrSO_4$(s)
c) SrCl(aq)

5. A pure yellow solid, lead(II) iodide, precipitates when two solutions are mixed. What are two reagents that could have produced this result. Write the balanced NET IONIC equation. List other reagents that can produce the same precipitate.

6. Write the balanced chemical equations from the following observations:
   a. A solution of phosphoric acid is added to an aqueous solution of barium nitrate. A white solid precipitate is observed.
   b. Solid diphosphorus pentoxide and gaseous sulfur dioxide are produced from the combustion of solid P_4S_5 .
   c. Solid magnesium nitride reacts with water to produce solid magnesium hydroxide and gaseous ammonia .

7. Write the balanced *net ionic equation* for each of the following:
   a) An aqueous solution of sodium iodide is treated with an aqueous solution of chlorine. Initially the solution was colorless. In the course of the reaction an orange color was observed.
   b) An aqueous solution of nitric acid is added to an aqueous solution of potassium hydroxide.
   c) A solution of potassium carbonate is added to a solution of iron (III) nitrate. A precipitate is formed.

8. When a piece of copper is placed in a solution of silver nitrate, a gray coating appears instantly on the surface of the copper. Write the balanced net ionic equation. Is the reaction redox? If so, identify the reduction agent. Are any spectator ions present? If so, identify.

9. Select an oxide from Se_2O, CaO, and P_4O_{10} which produces a basic solution when dissolved in water. Explain how the basic solution is formed.

10. Is the reaction below a redox reaction? Explain fully why is it or is not.
    $$2\ KClO_3 \rightarrow\ 2\ KCl\ +\ 3\,O_2$$

Key Concepts for Free Response:

1. If a precipitate forms with the H_2SO_4 and not with the NaCl, barium ions are present. The precipitate is $BaSO_4$. If a precipitate forms with NaCl but not with H_2SO_4, silver ions are present and the precipitate is AgCl. If a precipitate forms with both H_2SO_4 and NaCl, both ions were present in the contaminated water.

2. The precipitate is $BaSO_4$. Remaining in solution are the ions H^+ and Cl^- which will conduct an electric current.

3. In the solubility studies with water, three form aqueous solutions and one doesn't which is $BaCO_3$. In the reaction with HCl, both Na_2CO_3 and $BaCO_3$ form bubbles. Now the two carbonates have been identified. The two solids which formed a solution with water and did not react with HCl can now be identified by mixing with Na_2CO_3. The $BaCl_2$ solution will form a precipitate and the NaCl solution will not.

4: a) NaOH(aq) and HCl(aq) will react to form a pure aqueous solution of NaCl
    b) $Sr(NO_3)_2$(aq) and Na_2SO_4(aq) will reaction to form the ppt $SrSO_4$ and a solution containing the other ions. The pure $SrSO_4$ can be filtered and dried.
    c) $SrCO_3$(s) is treated with HCl(aq) to form pure SrCl(aq) after the CO_2(g) has escaped.

5. If you mix solutions of lead(II) nitrate and sodium iodide, lead(II) iodide will precipitate and all other ions will be in solution. $Pb^{2+}{}_{(aq)} + 2\,I^-{}_{(aq)} \rightarrow PbI_{2(s)}$ Other combinations that will produce the same precipitate are lead (II) acetate and potassium iodide. Any soluble lead (II) compound and any soluble iodide compound will produce this result.

6. a) $2\,H_3PO_4$(aq) $+ 3\,Ba(NO_3)_2$ (aq) $\rightarrow Ba_3(PO_4)_2$(s) $+ 6\,HNO_3$(aq)
    b) P_4S_5 (s) $+ 10\,O_2$ (g) $\rightarrow$ $2\,P_2O_5$ (s) $+$ $5\,SO_2$(s)
    c) Mg_3N_2 (s) $+ 6\,H_2O$ (l) $\rightarrow 3\,Mg(OH)_2$(s) $+ 2\,NH_3$(g)

7. a) $2\,I^-$(aq) $+ Cl_2$(aq) $\rightarrow 2\,Cl^-$(aq) $+ I_2$(aq)
    b) H^+(aq) $+ OH^-$(aq) $\rightarrow H_2O$ (l)
    c) $2\,CO_3{}^{2-}$(aq) $+ 3\,Fe^{3+}$(aq) $\rightarrow Fe_2(CO_3)_3$(s)

8. $2\,Cu$ (s) $+ Ag^+$(aq) $\rightarrow Ag$ (s) $+ Cu^{2+}$(aq)
    Copper is oxidized and silver is reduced. The reducing agent is the copper. The nitrate ions do not play a role in the reaction, they are spectator ions.

9. Metallic oxides can form basic solutions such as
    CaO (s) $+ H_2O$ (l) $\rightarrow$ Ca^{2+} (aq) $+ 2\,OH^-$(l)

10. Yes it is redox. The oxygen is oxidized from -2 to 0 while the chlorine is reduced from an oxidation number of +5 to -1. Redox reactions involve a transfer of electrons.

Section B: Multiple Choice

11. In a balanced chemical equation, what is balanced?

    a) atoms   b) moles   c) molecules

    d) atoms and molecules   e) moles and atoms

12. Phosphorus trichloride may be prepared by the reaction of phosphorus with chlorine gas according to the equation below.

$$\underline{\hspace{3em}}P_4 + \underline{\hspace{3em}}Cl_2 \rightarrow \underline{\hspace{2em}} PCl_3$$

Which of the following is the most correct set of stoichiometric coefficients to balance this equation?

    a) 2, 6, 8   b) 1, 3, 4   c) 2, 3, 2

    d) 1, 6, 4   e) 3, 9, 3

13. Naturally occurring sulfide ores can be "roasted" by heating in air. Balance the equation for the roasting of copper (I) sulfide with oxygen gas to produce copper and sulfur dioxide.

$$\underline{\hspace{3em}} Cu_2S(s) \; + \; \underline{\hspace{3em}} O_2(g) \; \rightarrow \; \underline{\hspace{2em}} Cu\,(s) \; + \; \underline{\hspace{2em}}SO_2(g)$$

Which of the following is the most correct set of stoichiometric coefficients to balance this equation?

    a) 1, 1, 2, 1   b) 2, 4, 2, 2   c) 1, 1, 1, 1

    d) 2, 2, 2, 2   e) 2, 1, 4, 2

14. Carbon disulfide, CS_2, can be made from the reaction of graphite C and SO_2 which also produces carbon monoxide.

$$\underline{\hspace{3em}} C \; + \; \underline{\hspace{3em}} SO_2 \rightarrow \underline{\hspace{2em}} CS_2 + \underline{\hspace{3em}} CO$$

Which of the following is the most correct set of stoichiometric coefficients to balance this equation?

    a) 2, 1, 1, 2   b) 5, 2, 3, 2   c) 4, 2, 1, 4

    d) 3, 3, 1, 2   e) 5, 2, 1, 4

15. When glucose undergoes complete combustion, the products are carbon dioxide and water.

_____ $C_6H_{12}O_6$ + _____ O_2 → _____ CO_2 + _____ H_2O

Which of the following is the most correct set of stoichiometric coefficients to balance this equation?

a) 1, 9, 6, 6   b) 1, 6, 6, 6   c) 2, 12, 6, 12

d) 2, 12, 12, 12   e) 1, 9, 12, 12

16. Ammonia will react with fluorine in the gaseous state to produce dinitrogen tetrafluoride and hydrogen fluoride.

_____ $NH_3(g)$ + _____ $F_2(g)$ → _____ $N_2F_4(g)$ + _____ $HF(g)$

Which of the following is the most correct set of stoichiometric coefficients to balance this equation?

a) 2, 1, 1, 6   b) 2, 3, 1, 6   c) 2, 5, 1, 6

d) 2, 10, 1, 6   e) 2, 5, 1, 6

17. Write and balance the equation for the combustion of butane, C_4H_{10}, with oxygen.

a) C_4H_{10} +   13 O_2 →   4 CO_2 + 5 H_2O

b) 2 C_4H_{10} +   13 O_2 → 8 CO_2 + 10 H_2O

c) 2 C_4H_{10} +   23 O_2 → 8 CO_2 + 5 H_2O

d) C_4H_{10} +   16 O_2 → 4 CO_2 + 10 H_2O

e) C_4H_{10} +   23 O_2 → 16 CO_2 + 5 H_2O

18. Write and balance the equation for the complete combustion of cyclohexane, C_6H_{12}, with oxygen.

a) C_6H_{12} +   18 O_2 →   6 CO_2 + 6 H_2O

b) C_6H_{12} +   9 O_2 →   6 CO_2 + 6 H_2O

c) C_6H_{12} +   6 O_2 →   6 CO_2 + 6 H_2

d) C_6H_{12} +   6 O_2 →   6 CO + 6 H_2O

e) C_6H_{12} +   12 O_2 →   6 CO_2 + 6 H_2

19. Write and balance the equation for the reaction of nitric acid with solid sodium carbonate.
    a) $H_2NO_3(aq) + NaCO_3(s) \rightarrow H_2O(l) + CO_2(g) + NaNO_3(aq)$
    b) $2\,HNO_3(aq) + Na_2CO_3(s) \rightarrow H_2(g) + CO_2(g) + Na_2NO_3(aq)$
    c) $2\,HNO_3(aq) + Na_2CO_3(s) \rightarrow H_2O(l) + CO_2(g) + 2\,NaNO_3(aq)$
    d) $2\,HNO_3(aq) + Na_2CO_3(s) \rightarrow H_2(g) + CO_2(g) + 2\,NaNO_3(aq)$
    e) $2\,HNO_3(aq) + NaCO_3(s) \rightarrow H_2(g) + CO_2(g) + NaNO_3(aq)$

20. Which equation below best represents the *balanced, net ionic equation* for the reaction of a solution of magnesium hydroxide with a solution of hydrochloric acid.
    a) $Mg^{2+}(aq) + 2\,Cl^-(aq) \rightarrow MgCl_2(s)$
    b) $Mg^{2+}(aq) + 2\,Cl^-(aq) \rightarrow MgCl_2(aq)$
    c) $2\,OH^-(aq) + 2\,Cl^-(aq) \rightarrow H_2O(l) + OCl_2(aq)$
    d) $Mg(OH)_2(aq) + 2\,Cl^-(aq) \rightarrow MgCl_2(s) + 2\,OH^-(aq)$
    e) $OH^-(aq) + H^+(aq) \rightarrow H_2O(l)$

21. Which equation below best represents the *balanced, net ionic equation* for the reaction of a solution of barium nitrate with a solution of potassium carbonate.
    a) $Ba^{2+}(aq) + CO_3^{2-}(aq) \rightarrow BaCO_3(s)$
    b) $K^+(aq) + NO_3^-(aq) \rightarrow KNO_3(s)$
    c) $Ba^{2+}(aq) + NO_3^-(aq) \rightarrow K_2CO_3(aq) + Ba^{2+}(s)$
    d) $Ba(NO_3)_2(aq) + K^+(aq) \rightarrow KNO_3(s) + Ba^{2+}(aq)$
    e) $Ba(NO_3)_2(aq) + CO_3^{2-}(aq) \rightarrow BaCO_3(s) + 2\,N_2(g) + 3\,O_2(g)$

22. Iron metal reacts with oxygen gas to form iron (III) oxide, a solid compound. Write the balanced equation.
    a) $2\,Fe(s) + O_2(g) \rightarrow Fe_2O_3(s)$
    b) $3\,Fe(s) + O_2(g) \rightarrow Fe_3O_2(s)$
    c) $Fe(s) + 3\,O_2(g) \rightarrow FeO_3(s)$
    d) $2\,Fe(s) + 3\,O_2(g) \rightarrow 2\,FeO_3(s)$
    e) $4\,Fe(s) + 3\,O_2(g) \rightarrow 2\,Fe_2O_3(s)$

23. Chlorine gas is bubbled into a solution of potassium iodide which is colorless. After about 3 minutes a red color is observed in the solution. Write the *balanced net ionic* equation.

a) $Cl_2(g) + 2 I^-(aq) \rightarrow I_2 (aq) + 2 Cl^-(aq)$

b) $Cl_2(g) + 2 K^+(aq) \rightarrow 2 K(s) + 2 Cl^-(aq)$

c) $2 Cl_2(g) + I_2^-(aq) \rightarrow I_2 (aq) + 2 Cl_2^-(aq)$

d) $2 Cl^-(aq) + 2K^+(aq) \rightarrow 2 K(s) + Cl_2 (aq)$

e) $Cl_2(g) + 2 KI(aq) \rightarrow 2K^+(aq) + I_2 (aq) + Cl_2^-(aq)$

24. Solutions of barium nitrate and ammonium sulfate are mixed. A white precipitate is observed. Write the *balanced net ionic* equation.

a) $Ba^{2+}(aq) + SO_4^{2-}(aq) \rightarrow BaSO_4 (s)$

b) $2 NH_4^+(aq) + SO_4^{2-}(aq) \rightarrow (NH_4)_2SO_4 (s)$

c) $NH_4^+(aq) + NO_3^-(aq) \rightarrow NH_4NO_3(s)$

d) $Ba^{2+}(aq) + 2 NH_4^+(aq) + 6 H_2O(l) \rightarrow Ba(NO_3)_2(s) + 20 H^+$

b) $Ba^{2+}(aq) + 2 NH_4^+(aq) + 2 SO_4^{2-}(aq) \rightarrow (NH_4)_2SO_4 (s) + BaSO_4(s)$

25. A solution of sulfuric acid is added to a solution of sodium hydroxide. Write the *balanced net ionic* equation.

a) $H_2SO_4 (aq) + 2 NaOH(aq) \rightarrow H_2O(l) + Na_2SO_4(s)$

b) $SO_4^{-2}(aq) + 2 Na^+(aq) \rightarrow Na_2SO_4(aq)$

c) $SO_4^{-2}(aq) + 2 H^+(aq) \rightarrow H_2SO_4(aq)$

d) $H_2 (aq) + 2 OH^-(aq) \rightarrow 2 H_2O(l)$

e) $H^+(aq) + OH^-(aq) \rightarrow H_2O(l)$

26. A solution of sodium carbonate is treated with a solution of nitric acid. Bubbles are observed. There is no evidence of a precipitate. Write the *balanced* equation.

a) $Na_2CO_3 (aq) + 2 HNO_3 (aq) \rightarrow H_2O(l) + CO_2(g) + 2 NaNO_3(aq)$

b) $Na_2CO_3 (aq) + 2 HNO_3 (aq) \rightarrow H_2CO_3(aq) + 2 NaNO_3(aq)$

c) $NaCO_3 (aq) + 2 HNO_3 (aq) \rightarrow$
$$H_2(g) + CO_2(g) + 3 O_2(g) + N_2(g) + Na_2O (aq)$$

d) $2 NaCO_3 (aq) + 2 HNO_3 (aq) \rightarrow H_2O(l) + 2 CO(g) + 3 O_2(g) + NaNO_3(aq)$

e) $2 NaCO_3 (aq) + 2 HNO_3 (aq) \rightarrow H_2O(l) + 2 CO_2(g) + N_2(g) + 2 NaNO_3(aq)$

27. If solutions of the same concentration are prepared with the following substances, which one will have the greatest electrical conductivity?
    a) CH_3CO_2H b) HCl c) $NaCl$
    d) $CaCl_2$ e) Cl_2

28. The classification of the following reactions *in order* is
$$HCl(g) \ + \ NH_3(g) \ \rightarrow \ NH_4Cl$$
$$2\,HgO(s) \ \rightarrow \ O_2\,(g) \ + \ 2\,Hg(l)$$
$$HCl(aq) \ + \ AgNO_3(aq) \ \rightarrow \ AgCl(s) \ + \ HNO_3(aq)$$
    a) acid-base, precipitation, and redox respectively.
    b) precipitation, acid-base, and redox respectively.
    c) redox, precipitation, acid-base respectively.
    d) acid-base, redox and precipitation respectively.
    e) redox, acid-base, precipitation respectively.

29. A precipitate will form when an aqueous solution of lead (II) nitrate is added to an aqueous solution of
    a) NH_4NO_3 b) $Mg(NO_3)_2$ c) $NaNO_3$
    d) KNO_3 e) $NaCl$

30. Which of the following is a weak acid?
    a) NH_3 b) HCl c) $HClO_4$
    d) CH_3CO_2H e) HNO_3

31. A solution of nitric acid contains which of the following ions in easily measurable quantities?
    a) H^+, N_2^-, O_2^- b) H^+, NO_2^- c) H_2^+, NO_3^-
    d) H_2^+, $2\,NO_2^-$ e) H^+, NO_3^-

32. Identify the spectator ion or ions (if any) in the redox reaction of a solution of lead(II) nitrate with zinc metal.
    a) Pb^{2+} b) Zn^{2+} c) NO_3^-
    d) H^+ and Pb^{2+} e) no spectator ions are present

33. The driving force for the reaction of zinc metal with a solution of lead(II) nitrate is
    a) the formation of a precipitate
    b) the formation of a gas
    c) the evolution of a gas
    d) the dissolving of a solid
    e) the transfer of electrons

34. When a solution of sodium hydroxide is mixed with a solution of potassium phosphate
    a) a precipitate will be observed   b) a gas will be evolved
    c) water will be formed   d) no reaction will be observed
    e) both a gas and a precipitate will be formed

35. When a solution of sodium chloride and a solution of ammonium nitrate are mixed
    a) $NH_4Cl(s)$ forms.   b) $NaNO_3 (s)$ forms.   c) $NaNH_4(s)$ forms.
    d) N_2 and O_2 gases are released.
    e) neither a precipitate nor a gas is formed.

36. Consider the equation: $2 NaI(aq) + Cl_2 (g) \rightarrow I_2 (aq) + 2 NaCl(aq)$
    The element undergoing reduction is
    a) sodium   b) iodide   c) chlorine
    d) iodine   e) water

37. Consider the equation: $I^-(aq) + ClO^-(aq) \rightarrow IO^-(aq) + Cl^-(aq)$
    The oxidizing agent is
    a) I^- b) ClO^- c) IO^-
    d) Cl^- e) H_2O

38. What species is the oxidizing agent in the reaction below?
                  $Cl_2 (aq) + 2 Br^-(aq) \longrightarrow 2 Cl^-(aq) + Br_2 (aq)$
    a) Cl_2 b) Br^- c) Cl^-
    d) Br_2 e) H_2O

39. The oxidation number of chromium in Na_2CrO_4 is
        a) –2   b) +2   c) +6   d) – 6   e) +8

40. The reducing agent (if any) in the following equation is

$$2 \, Mg(s) + \quad TiCl_4 \, (l) \quad \rightarrow \quad Ti \, (s) \quad + \quad 2 \, MgCl_2(s)$$

    a) $Mg(s)$ b) $TiCl_4$ (l)   c) Ti (s)

    d) $MgCl_2(s)$ e) not a redox reaction

41. The oxidation number of chlorine in $KClO_3$ is

    a)+6   b) +5   c) –1   d) –2   e) +2

42. The oxidation number of sulfur in $S_2O_3{}^{2-}$ is

    a) +8   b) +6   c) +4   d) +3   e) +2

43 . The oxidation number of manganese in $KMnO_4$ is

    a) –2   b) +3   c) +6   d) +7 e) +8

44 . The oxidation number of nitrogen in N_2O_5 is

    a) –3   b) +2   c) +3   d) +5 e) +10

45. Which of the following reactions represent an oxidation-reduction?

    i) $H_2SO_{3(aq)} \rightarrow H_2O_{(l)} + SO_{2(g)}$

    ii) $Zn_{(s)} + Cu(NO_3)_{2(aq)} \rightarrow Cu_{(s)} + Zn(NO_3)_{2(aq)}$

    iii) $Zn_{(s)} + S_{(s)} \rightarrow ZnS_{(s)}$

    a) i   b) i and ii   c) ii and iii

    d) i and iii   e) ii, ii and iii

46. Which of the following reactions represent an oxidation-reduction?

    i) $3 \, Mg \, (s) + N_2 \, (g) \rightarrow Mg_3N_2 \, (s)$

    ii) $H_2CO_{3(aq)} \rightarrow H_2O_{(l)} + CO_{2(g)}$

    iii) $Sr(NO_3)_{2(aq)} + Na_2SO_{4(aq)} \rightarrow SrSO_{4 \, (s)} + 2 \, NaNO_{3(aq)}$

    a) i   b) ii and iii   c) i and iii

    d) i and ii   e) i, ii, and iii

47. If lead(II) nitrate and sodium chloride solutions are mixed, what is it formula of the precipitate formed (if any)?

   a) $PbCl_4$ b) $PbCl_2$ c) $NaNO_3$

   d) Na_2Pb e) no precipitate is formed.

48. A white solid is either NaCl or $NaNO_3$. If an aqueous solution is prepared, which reagent will allow you to distinguish between the two compounds.

   a) H_2SO_4 b) HCl   c) $AgNO_3$

   d) $(NH_4)_2 SO_4$ e) H_2O

49. The solution which results from the reaction NaOH(aq) and HCl (aq) is the same as the result of the reaction of

   a) $Pb(NO_3)_2$(aq) and NaCl (aq)   b) KOH (aq) and HCl(aq)

   c) NaCl(aq) and $AgNO_3$(aq)   d) Na_2CO_3(aq) and HCl(aq)

   e) HCl (aq) and $(NH_4)_2SO_4$(aq)

50. The compound $Fe(OH)_3$ may be formed in the laboratory by

   a) the reaction of Fe_2O_3 (s) and NaCl (aq)

   b) the reaction of Fe_2O_3 (s) and HCl (aq)

   c) mixing solutions of $FeSO_4$ and NaOH

   d) mixing solutions of from Fe_2Cl_5 and H_2O

   e) mixing solutions of $FeCl_3$ and NaOH

51. When an aqueous solution of lead (II) nitrate is treated with an aqueous solution of potassium carbonate, one may observe

   a) the formation of a precipitate, $PbCO_3$

   b) the formation of a gas, CO_2

   c) both the formation of $PbCO_3$ precipitate and CO_2 gas

   d) the formation of two precipitates, KNO_3 and $PbCO_3$

   e) no reaction

52. Which of the following compounds is an electrolyte?

   a) table salt   b) sand   c) table sugar

   d) antifreeze   e) cooking oil

53. Which pairs of reagents (if any) could be used in an aqueous solution to prepare
    PURE manganese (II) sulfide by a precipitation reaction?
    a) $MnCO_3$ and Ag_2S b) $MnCl_2$ and Na_2S
    c) $MnCl_2$ and Na_2SO_4 d) $MnSO_4$ and PbS
     e) none of these reagents will produce pure manganese (II) sulfide

54. Which of these acids will dissociate 100%?
    a) HCO_2H b) CH_3CO_2H c) H_2CO_3 d) H_3BO_3 e) HNO_3

55. Which of the following reactions is NOT a redox reaction?
    a) $2 HgO(s) \rightarrow 2 Hg(l) + O_2(g)$
    b) $H_2(g) + Br_2(g) \rightarrow 2 HBr(g)$
    c) $H_2CO_3(aq) \rightarrow CO_2(g) + H_2O(l)$
    d) $2 HCl(aq) + Zn(s) \rightarrow H_2(g) + ZnCl_2(aq)$
    e) $2 NaI(aq) + Cl_2(aq) \rightarrow I_2(aq) + 2 NaCl(aq)$

Chapter 4: Multiple Choice Answers

11. a	21. a	31. e
12. d	22. e	32. c
13. a	23. a	33. e
14. e	24. a	34. d
15. a	25. e	35. e
16. c	26. a	36. c
17. b	27. d	37. b
18. b	28. d	38. a
19. c	29. e	39. c
20. e	30. d	40. a

41. b	51. a
42. e	52. a
43. d	53. b
44. d	54. e
45. c	55. c
46. c	
47. b	
48. c	
49. d	
50. e	

Chapter 5: Stoichiometry

Section A: Free Response

1. Consider the reaction of substance A with substance BC which produces substances AB_2 and C according to the equation below. If 6 moles of A and 8 moles of BC are allowed to react, how many moles (if any) of each species are present after the reaction. Explain your logic.

$$A \quad + \quad 2\,BC \quad \rightarrow \quad AB_2 \quad + \quad 2\,C$$

Reagents	A	BC	AB_2	C
Moles Present Before Reaction	6	8	0	0
Moles Present After Reaction				

2. Suggest some reasons why a chemist would design an experiment with a limiting reagent rather than perfoming it with the exact stoichiometric ratios.

3. Explain how you would prepare 50.0 mL of a 0.267 M NaCl solution. Include in your explanation the reagents and equipment you would need as well as the procedure you would follow.

4. A student prepared a solution by dissolving 10.0 grams of sodium sulfate in enough water to make 250. mL of solution. Does this solution contain enough sulfate ions to precipitate all the barium ions present in 5.00 mL of a 0. 384 M barium nitrate solution? Explain.

5. The density of liquid pentane, C_5H_{12}, is 0.626 g/mL at room temperature. How many grams of CO_2 can be produced when 25.0 milliliters of pentane measured at room temperature is burned?

6.   A chemist working in a municipal water treatment plant *always* adds an excess of sodium sulfate to the filtering equipment which carries out the reaction below. Briefly, explain why this is a good procedure.

$$Pb^{2+}\,(aq) + Na_2SO_4(aq) \longrightarrow PbSO_4(s) + 2\,Na^+(aq)$$

7. A chemical laboratory uses the equipment in the figure to determine the empirical formula of
    a sample.   The unknown is combusted in an oxygen atmosphere in Chamber A. The gaseous
    products flow to a carbon dioxide analyzer (Chamber B) and then to a water vapor analyzer
    (Chamber C). Using the laboratory data below determine the empirical formula of each
    compound.

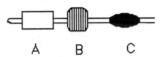

A   B   C

	Unknown X	Unknown Y	Unknown Z
Elements present	C, H	C, H, O	C, H, N
Mass of sample	7.22 grams	2.074 grams	6.50 grams
Mass of CO_2	22.6 grams	3.800 grams	15.9 grams
Mass of H_2O	9.27 grams	1.040 grams	4.33 grams

8. An orange solid with the percent composition below is heated. A violent reaction occurs
    producing a green solid with the composition given, a gaseous element, and a non-hazardous
    molecule in the vapor state. Write the balanced equation.

    Orange Solid: nitrogen   11.11%   Green solid: chromium   68.43%
                       hydrogen   3.20%   oxygen   31.57%
                       chromium   41.26%
                       oxygen   44.42%

9. If 2.75 grams of NaI are produced from a mixture initially containing 5.00 g I_2 and 1.00 g
    NaOH, what is the percent yield?
    $$3 I_2 + 6 NaOH \longrightarrow 5 NaI + NaIO_3 + 3 H_2O$$

10. Phosphine gas (PH_3) can be prepared in the two-step process:

    $$6 Ca(s) + P_4(s) \longrightarrow 2 Ca_3 P_2 (s)$$
    $$Ca_3 P_2 (s) + 6 H_2O \longrightarrow 2 PH_3(g) + 2 Ca(OH)_2 (s)$$

    If your equipment operates at 85.0% effiency, how many grams of phosphorus are needed
    to generated 500. grams of PH_3?

Answers to Free Response:

1.

Reagents	A	BC	AB_2	C
Moles Present Before Reaction	6	8 limiting	0	0
Moles Present After Reaction	6-4 = 2 2	0	4	8

2. If one reagent is only available in small amounts, expensive, hazardous, difficult to dispose, or requires specialized equipment, the chemist would design the experiment to consume all of that reagent. Sometimes reactions do not procede at 100% yield so if the concentrations of the reagents are controlled, the chances for a reaction may be increased the presence of one reagent in excess so that the limiting reagent will then be totally consumed

3. Weigh out 0.780 grams of sodium chloride crystals. Add to a 50.0 mL volumetric flask. Add a small amount of distilled water. Shake to dissolve. Then add water to the line of the volumetric flask. A wash bottle with distilled water will allow precise control for the final few drops. Shake to mix. The solution is now 0.267 M.

4. Yes, the solution contains 0.0704 moles SO_4^{2-} which is about a 36 fold excess to the 0.00192 moles Ba^{2+} in solution. All of the Ba^{2+} will precipitate as $BaSO_4$.

5. $C_5H_{12}(g) + 8 O_2 (g) \quad ----> 5 CO_2(g) + 6 H_2O (g)$
   47.7 grams CO_2

6. Using an excess of sodium sulfate guarantess that all of the Pb^{2+} will be removed from the water. In a municipal water supply, the levels of Pb^{2+} must be kept as low as possible to avert lead poisoning. Thus the excess sodium sulfate

7. Unknown X is determined from the molar ratios of carbon and hydrogen in the masses of carbon dioxide and water respectively. Empirical formula: CH_2
   Unknown Y contains oxygen which cannot be measured directly. The carbon and hydrogen are analyzed from molar ratios in CO_2 and H_2O while the oxygen is determined by substracting from the sample mass. Empirical formula: $C_3H_4O_2$
   Similarly, Unknown Z is C_3H_4N

8. $(NH_4)_2Cr_2O_7 (s) \quad \rightarrow \quad Cr_2O_3(s) + N_2 (g) + 4 H_2O (g)$

9. The limiting reagent is NaOH. The yield is 88.1% .

10. To generate 500 g phosphine with this equipment 536 g of phosphorus (P_4) are needed.

Section B: Multiple Choice

11. How many grams of carbon are needed to react completely with 75.2 grams of SiO_2 according to the following equation?

$$SiO_2(s) \ + \ 3\,C(s) \ \rightarrow SiC(s) \ + 2\,CO(g)$$

a) 15.0 g   b) 61.5 g   c) 20.5 g

d) 32.8   e) 45.1 g

12. How many milliliters of bromine (density = 3.19 g/mL) are needed to react completely with 85.0 grams of NH_3 to produce ammonium bromine and nitrogen according to the equation below ?

$$3\,Br_2(l) + \ 8\,NH_3(g) \ \rightarrow 6\,NH_4Br(s) \ + N_2(g)$$

a) 46.9 mL   b) 93.7 mL   c) 249.9 mL

d) 298.9 mL   e) 315.4 mL

13. How many grams of aluminum oxide are needed to prepare 310.0 grams of aluminum according to the equation below?

$$Al_2O_3\,(s) \ + \ 3\ C(s) \ \xrightarrow{\ electricity\ } 2\,Al(s) \ + \ 3\,CO(g)$$

a) 292.9 g   b) 430.8 g   c) 585.8 g

d) 1171 g   e) 1282 g

14. If 150.0 g SiO_2 and 60.0 grams C are allowed to react according to the equation below, what is the maximum amount of CO that can be produced? Answer in grams.

$$SiO_2(s) \ + \ 3\,C_{(s)} \ \rightarrow SiC_{(s)} \ + 2\,CO_{(g)}$$

a) 93.4 g   b) 70.0 g   c) 140. g

d) 210. g   e) 245 g

15. If 35.0 grams of bromine and 20.0 grams of ammonia are allowed to react according to the equation below, what is the maximum number of grams of N_2 that can be produced?

$$3\,Br_2(l) + \ 8\,NH_3(g) \ \rightarrow \ 6\,NH_4Br(s) \ + N_2(g)$$

a) 2.05 g   b) 5.15 g   c) 6.13 g

d) 41.2 g   e) 60.5 g

16. If 25.0 grams of Al_2O_3 and 75.0 grams of carbon react according to the equation below, what is the maximum number of grams of Al that can be produced?

$$Al_2O_3 + 3 C \xrightarrow{\text{electricity}} 2 Al + 3 CO$$

a) 112 grams   b) 6.61 grams   c) 112 grams

d) 100.0 grams   e) 13.2 grams

17. The following reaction occurs in the head of a match.

$$P_4S_3 (s) + 8 O_2 (g) \rightarrow P_4O_{10}(s) + 3 SO_2 (g)$$

How many grams of O_2 are needed to react with 0.450 grams of P_4S_3 ?

a) 0.0654 g   b) 0.261 g   c) 0.523 g

d) 3.60 g   e) 1.83 g

18. Given the unbalanced equation: $Fe + O_2 \rightarrow Fe_2O_3$

The number of moles of oxygen gas which will react with 1.6 mole of iron to produce iron (III) oxide is

a) 1.2 mol   b) 1.6 mol   c) 2.2 mol

d) 2.4 mol   e) 3.2 mol

19. Hydrogen gas burns to give water in the following reaction

$$2 H_2 + O_2 \rightarrow 2 H_2O$$

If 4.0 g of hydrogen gas and 8.0 g of oxygen gas are available as reactants, what is the limiting reagent, if any?

a) H_2 b) O_2 c) heat

d) H_2O e) there is no limiting reagent

20. Given the balanced reaction: $A + 3B \rightarrow 2 C$.

The molar mass of C is 20.0 grams. If one *mole* of A produces 10.0 *grams* of C, what is the percent yield of the reaction?

a) 400%   b) 50%   c) 25%

d) 10%   e) 1.0%

21. How many grams of carbon dioxide are produced by the complete combustion of 5.00 grams of acetylene (C_2H_2)?

a) 8.45 g   b) 20.0 g   c) 33.8 g

d) 12.6 g   e) 16.9 g

22. How many grams of carbon dioxide are produced by the complete combustion of 5.00 grams of pentane (C_5H_{12})?

    a) 0.346 g   b) 3.04 g   c) 4.15 g

    d) 15.2 g   e) 12.7 g

23. How many *moles* of magnesium (if any) remain when 5.00 grams of magnesium is burned in 2.50 grams of pure oxygen according to the equation?

$$2\,Mg\,(s)\ +\ O_2\,(g)\ \rightarrow\ 2\,MgO(s)$$

    a) 0.206 mol   b) 0.0497 mol   c) 0.0781 mol

    d) 0.156 mol   e) zero, it is totally consumed

24. When 10.0 grams of mercury (II) oxide was decomposed, a student obtained 5.00 grams of mercury. What was the per cent yield?

$$2\,HgO(s)\ \rightarrow\ 2\,Hg(l)\ +\ O_2(g)$$

    a) 46.0%   b) 50.0%   c) 54.0 %

    d) 33.3%   e) 10.0%

25. How many grams of K_2SO_4 (1 mole weighs 174.2 grams) needed to prepare 65.0 mL of a 0.658 M solution?

    a) 5.78 g   b) 12.3 g   c)17.3 g

    d) 24.6 g   e) 7.45 g

26. If you dissolve 8.14 grams of calcium chloride, $CaCl_2$, in enough water to make 125.0 mL of solution, what is the molarity of the solution?

    a) 0.00916 M   b) 5.84 x 10-4 M   c) 0.586 M

    d) 1.71 M   e) 2.24 M

27. How many milliliters of 0.515 M $Ba(NO_3)_2$ solution will provide 1.25 grams of barium nitrate?

    a) 2.43 mL   b) 4.78 mL   c) 5.07 mL

    d) 7.34 mL   e) 9.29 mL

28. When a 75.5 gram sample of sulfur-containing coal was burned 4.30 grams of SO_2 was analyzed. Assuming the coal to be a mixture of carbon and sulfur, what is the % sulfur in the coal?

a) 1.00%   b) 1.43%   c) 1.90%
d) 5.70%   e) 7.92 %

29. The reaction below was known to proceed at 85.3% yield of SO_2 .

$$P_4S_5 \text{ (s) } + \quad 5\,O_2 \text{ (g) } \rightarrow \quad 2\,P_2O_5 \text{ (s) } \quad + \quad 5\,SO_2(g)$$

How many grams of P_4S_5 must be burned to produce 50.0 grams of SO_2?

a) 10.4 g   b) 15.1 g   c) 44.3 g
d) 46.8 g   e) 52.0 g

30. When 5.00 grams of Na_2CO_3 was treated with excess HCl, 1.50 grams of CO_2 was obtained. What was the % yield?

a) 16.6 %   b) 23.1%   c) 72.3 %
d) 30.0 %   e) 34.1%

31. To what volume must 150 mL of a 3.60 M solution be diluted to prepare a solution which is 2.40 M?

a) 1296 mL   b) 1000 mL   c) 444 mL
d) 285 mL   e) 225 mL

32. How many milliliters of 6.00 M NaOH should be added to 120.0 mL of a 0.300M HNO_3 solution to completely neutralize the acid?

a) 6.00 mL   b) 12.0 mL   c) 120.0 mL
d) 600. mL   e) 2400.0 mL

33 . How many milliliters of 0.414 M $KMnO_4$ are required to react completely with 7.22 grams of solid iron under acidic conditions according to the equation below?

$$24\,H^+_{(aq)} + 5\,Fe_{(s)} + \quad 3\,MnO_4^-{}_{(aq)} \text{ -- } \rightarrow \quad 5\,Fe^{3+}{}_{(aq)} \quad + 3\,Mn^{2+}{}_{(aq)} \quad + 12\,H_2O_{(l)}$$

a) 77.5 mL   b) 97.7 mL   c) 187 mL
d) 387 mL   e) 421 mL

34. What is the concentration of NH_4^+ in a 0.150 M $(NH_4)_2SO_4$ solution?
    a) 0.150 M   b) 0.300 M   d) 0.600 M
    d) 0.222 M   e) 0.0750 M

35. Which solution will have the highest concentration of potassium ion?
    a) 3.0 M KCl   b) 1.5 M K_3PO_4 c) 2.5 M K_2SO_4
    d) 3.0 M KOH   e) 1.5 M KNO_3

36. Which of the following solutions will have the least electrical conductivity?
    a) 0.1 M $(NH_4)_3PO_4$ b) 0.1 M $BaCl_2$ c) 0.1 M Na_2SO_4
    d)   0.1 M $NaNO_3$ e) 0.1 M K_3PO_4

37. You have 500.0 mL of a 0.64 M NaCl solution. You take out 100.0 mL. The molarity of the
    remaining solution is
    a)   0.64 M   b) 0.51 M   c) 0.40 M
    d)   0.32M   e) 0.13 M

38. You add 4.00 mL of water to 6.00 mL of a 0.155 M NaCl solution. What is the new
    concentration?
    a) 9.3 x 10^{-3} M   b) 0.093 M   c) 0.107 M
    d) 1.7 x 10^{-3} M   e) 0.165 M

39. If you need 50.0 mL of a 0.250 M $KMnO_4$ solution which method do you use to prepare it?
    a) Dissolve 39.5 g $KMnO_4$ in enough water to make 50.0 mL solution.
    b) Dissolve 12.5 g $KMnO_4$ in enough water to amake 50.0 mL solution.
    c) Dissolve 1.98 g $KMnO_4$ in enough water to make 50.0 mL solution.
    d) Dilute 20.0 mL of 0.500 M $KMnO_4$ to 50.0 mL.
    e) Dilute 2.50 mL of 1.00 M $KMnO_4$ to 50.0 mL.

40. The molar mass of KBr is 119.0. How many grams of KBr are needed to prepare 50.0 mL
    of a 0.250 M solution?
    a) 1.05 g   b) 4.76 g   c) 29.75 g
    d) 0.0125 g   e) 1.49 g

41. A 2.80 gram sample of impure benzoic acid ($C_6H_5CO_2H$) requires 35.0 mL of 0.430 M
    NaOH for neutralization according to the equation below. What is the percentage of benzoic
    acid in the sample? Assume the impurity does not react or affect the neutralization reaction
    of the benzoic acid.

    $C_6H_5CO_2H(aq) + OH^-(aq) \rightarrow H_2O(l) + C_6H_5CO_2^-(aq)$

    a) 1.84%   b) 15.0%   c) 35.7 %

    d) 52.5%   e) 65.7%

42. A mixture of potassium chlorate ($KClO_3$) and potassium chloride (KCl) weighs 4.55 grams.
    When the sample is heated the $KClO_3$ decomposes according to the equation below but the
    KCl does not change. After completion of the reaction, the residue weighed 3.82 grams.
    What was the percent $KClO_3$ in the mixture?

    $2\ KClO_3(s) \rightarrow 2\ KCl(s) + 3\ O_2\ (g)$

    a) 40.9%   b) 83.9%   c) 25.7 %

    d) 1.86%   e) 16.0%

43. What is the empirical formula for a compound which is 75.92% carbon, 6.37% hydrogen
    and 17.71% nitrogen?

    a) C_5H_5N b) C_6H_6N c) $C_6H_6N_3$

    d) $C_7H_6N_2$ e) $C_6H_5N_1$

44. What is the empirical formula of a compound which is 40.72% antimony and 59.28%
    chlorine?

    a) Sb_2Cl_4 b) Sb_2Cl_5 c) $SbCl_2$

    d) $SbCl_3$ e) $SbCl_5$

45. What is the empirical formula of a compound which is 64.80% carbon, 6.35 % hydrogen
    and 28.83 % sulfur?
    a) C_5H_6S b) $C_6H_7S_2$ c) C_5H_7S
    d) C_6H_7S e) $C_5H_5S_2$

46. What is the empirical formula of a compound which is 57.10% carbon, 4.79 % hydrogen,
    and 38.11 % sulfur?
    a) C_2H_2S b) C_4H_4S c) $C_5H_5S_2$
    d) C_6H_6S e) C_5H_8S

47. What is the empirical formula of a compound which is 43.01% carbon, 2.58 % hydrogen
    and 54.41 % chlorine?
    a) $C_4H_3Cl_2$ b) $C_4H_3Cl_3$ c) $C_6H_5Cl_2$
    d) $C_7H_5Cl_3$ e) $C_6H_7Cl_2$

48. Titanium (IV) oxide may be treated with chloride and carbon to form titanium (IV) chloride
     with the release of carbon monoxide according to the following equation:
$$TiO_2(s) + 2\ Cl_2(g) + 2\ C(s) \rightarrow \quad TiCl_4(g) + 2\ CO(g)$$
     Suppose 50.0 grams of $TiO_2(s)$ is reacted with excess $Cl_2(g)$ and C(s) and 22.0 grams of
     CO is isolated. Calculate the percentage yield of CO.
     a) 17.5%   b) 35.1%   c) 62.7%
     d) 83.5%   e) 79.6%

49. Caffeine has the formula $C_8H_{10}N_4O_2$. If 5.00 milligrams of caffeine is burned, how many
    milligrams of CO_2 are produced?
    a) 1.13 mg   b) 1.76 mg   c) 2.06 mg
    d) 9.06 mg   e) 5.67 mg

50. A 2.44 gram sample of a compound containing only C, H, and N was burned in excess
     oxygen to yield 6.79 g CO_2 and 1.39 g H_2O. Calculate the empirical formula.
     a) C_3H_6N b) C_4H_4N c) C_5H_5N
     d) $C_3H_6N_2$ e) $C_3H_5N_2$

51. An unknown compound contains the elements carbon, hydrogen, and oxygen. When a 5.422 gram sample of a compound was analyzed by combustion in pure oxygen, 6.878 grams CO_2 and 1.872 grams H_2O were obtained. Calculate the empirical formula.

   a) $C_2H_4O_2$ b) $C_3H_4O_4$ c) $C_2H_6O_2$

   d) $C_2H_2O_3$ e) $C_2H_6O_3$

52. A 4.50 gram sample of a compound containing only carbon, hydrogen, and nitrogen was burned in excess oxygen to yield 12.76 g CO_2 and 2.22 g NO_2. The hydrogen-containing products were not identified. Calculate the % composition of C, N, and H in the sample.

   a) C_6H_7N b) C_3H_4N c) C_7H_6N

   d) $C_5H_6N_2$ e) C_6H_4N

53. For the reaction of 15.0 grams of magnesium and 5.00 grams of nitrogen according to the equation below, which statement is true?

$$3\,Mg \;+\; N_2 \;\;\rightarrow\;\; Mg_3N_2$$

   a) Mg is limiting and 0.439 mol N_2 is in excess

   b) N_2 is limiting and 0.082 mol Mg is in excess

   c) 20.0 grams of Mg_3N_2 is produced

   d) 15.0 grams of Mg_3N_2 is produced

   e) There is no limiting reagent.

54. What volume of 0.166 M NaOH is needed to neutralize 10.0 mL of 0.261M HNO_3?

   a) 0.0157 mL   b) 1.57 mL   c) 15.7 mL

   d) 157 mL   e) 1.57 L

55. If 18.5 mL of a NaOH solution is required to neutralize 25.0 mL of 0.457 M HCl, what is the molarity of the NaOH?

   a) 1.14×10^{-2} M   b) 8.45×10^{-3}M   c) 1.82 M

   d) 1.24 M   e) 0.617 M

56. What volume of 0.150 M NaOH is needed to react completely with 3.45 g iodine according to the equation $3\,I_2 \;+\; 6\,NaOH \;\;\rightarrow\;\; 5\,NaI \;+\; NaIO_3 \;+\; 3\,H_2O$

   a) 181 mL   b) 45.3 mL   c) 1.02 mL   d) 2.04 mL   e) 4.08 mL

Chapter 5 : Answers to Multiple Choice:

11. e	21. e	31. e
12. b	22. d	32. a
13. c	23. b	33. c
14. a	24. c	34. b
15. a	25. e	35. c
16. e	26. c	36. d
17. c	27. e	37. a
18. a	28. c	38. b
19. b	29. e	39. c
20. c	30. c	40. e

41. e	51. b
42. a	52. a
43. a	53. b
44. e	54. c
45. d	55. e
46. b	56. a
47. d	
48. e	
49. d	
50. c	

Chapter 6
Energy and Chemical Reactions

Section A: Free Response

1. Consider the reaction A → B which can occur by Pathway 1, a one step process or by Pathway 2 which consists of a series of intermediate steps. Is the enthalpy change in going from A to B by Pathway 2 greater than, less than, or the same as the enthalpy change by Pathway 1? Explain.

    Pathway 1:   A → B   Pathway 2: A → C
                                              C → D
                                              D → B

2. Using the following information:

    $$H_2(g) + F_2(g) \rightarrow 2\,HF(g) \qquad \Delta H = -537\ kJ$$
    $$C(s) + 2\,F_2(g) \rightarrow CF_4(g) \qquad \Delta H = -680.\ kJ$$
    $$2\,C(s) + 2\,H_2(g) \rightarrow C_2H_4(g) \qquad \Delta H = +52.\ kJ$$

    calculate the enthalpy for the reaction of ethylene, C_2H_4, with F_2 according to the equation below.

    $$C_2H_4(g) + 6\,F_2(g) \rightarrow 2\,CF_4(g) + 4\,HF(g)$$

3. Calculate the enthalpy of reaction for the combustion of 9.25 grams of butane, C_4H_{10}, according to the equation

    $$C_4H_{10}\,(l) + \frac{13}{2}\,O_2\,(g) \rightarrow 4\,CO_2(g) + 5\,H_2O(l)$$

    Standard Enthalpies of Formation (kJ/mol)
    $$CO_{2(g)} = -394;\quad H_2O_{(l)} = -286;\quad CH_3COOH_{(l)} = -484$$

4. A 40.0 g piece of metal at 203.0°C is dropped into 100.0 g of water at 25.0 °C. The water temperature rises to 33.0°C. Calculate the specific heat of the metal. Assume that all the heat lost by the metal is transferred to the water and no heat is lost to the surroundings.

5. Explain the first law of thermodynamics in a few words and in an equation. Explain how the positive and negative signs are related to each term.

6. Outline an experimental procedure to determine the specific heat of zinc.

7. Diagram the energy relationships for the reaction

   A + B + 35 kJ → C

   Label appropriately.

8. In a calorimeter 2.25 grams of acetic acid, CH_3COOH, is completely burned with the energy change in the reaction being recorded by a temperature change in a water "bath" surrounding the reaction as diagrammed. Given the data below, find the final temperature of the water bath if its initial temperature is 25.0°C and its mass is 550.0 grams. Assume no heat loss to the calorimeter.

   $\Delta H_f^{\circ}(CO_{2(g)}) = -394$ kJ /mol   Specific Heat of Water 4.184 J/g· K

   $\Delta H_f^{\circ}(H_2O_{(l)}) = -286$ kJ /mol

   $\Delta H_f^{\circ}(CH_3COOH_{(l)}) = -484$ kJ /mol

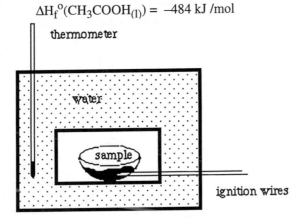

9. When 2.50 grams of oxalic acid, $H_2C_2O_4$, is burned in a calorimeter containing 100.00 grams of water, the temperature of the water increases 1.500 degrees. Calculate the heat of combustion for *one mole of oxalic acid* and express your answer in kJ/mol. Assume the reaction chamber absorbs no heat.

10. Which equation represents a reaction that could be used as a "cold pack" by athletes? Explain.

              Equation 1: A + B + 25 kJ → C
              Equation 2:   D + E → F + 45 kJ

Key Concepts for Free Response

1. The enthalpy change will be the same no matter the pathway. Enthalpy is a state function.

2. –2486 kJ

3. –337 kJ

4. 0.47 J/K · g

5. The first law of thermodynamics states that the change in energy of a process is equal to the amount of heat transferred and the amount of work performed through the equation, $\Delta E = q + w$. Heat energy or work leaving the system have negative values. If the heat transfer, q, is from the system to the surroundings, it is negative. If the work transfer, w, is from the system to the surroundings, it is negative. Positive values are for energy or work transferred from the surroundings to the system.

6. Weigh out and record an amount of water and record the temperature. Weigh out and record an amount of zinc. Heat the zinc and record the temperature. Drop into the water and measure the temperature change. The specific heat can be calculated by using the equation: $q_{zinc} + q_{water} = 0$.

7. Energy Diagram for A + B + 35 kJ $\longrightarrow$ C

Energy ↑
        C ▬
   A + B
        ▬

8. 39.3 °C

9. -22.6 kJ/mol

10. The first equation (endothermic) because it absorbs heat from the surroundings, which can be put to good use. The container of this reaction will feel cold.

Section B: Multiple Choice

11. How many calories are equivalent to 364 J?

    a) 1523 cal   b) 87.0 cal   c) 14.6 cal   d) 1.33 cal   e) 0.364 cal

12. How may joules are equivalent to 37.7 cal?

    a) 9.01 J   b) 9.43 J   c) 1.51 J   d) 4.184 J   e) 158 J

13. The quantity of heat that is needed to raise the temperature of a sample of a substance 1.00 kelvin is called its

    a) heat capacity   b) specific heat capacity   c) enthalpy

    d) calorimetry   e) kinetic energy

14. Specific heat capacity is

    a) amount of heat energy needed to change 1.00 g of substance by 1.00 K.

    b) amount of heat energy needed to change 1.00 mol of substance by 1.00 K.

    c) amount of energy required to melt 1.00 g of substance.

    d) amount of substance that is heated by 1.00 K.

    e) the number of kelvins that 1.00 g of substance is raised by heating it for 1.00 minute.

15. Equal masses of two substances, A and B, each absorb 25 Joules of energy. If the temperature of A increases by 4 degrees and the temperature of B increases by 8 degrees, one can say that

    a) the specific heat of A is double that of B.

    b) the specific heat of B is double that of A.

    c) the specific heat of B is negative.

    d) the specific heat of A is negative.

    e) the specific heat of B is triple that of A.

16. The same amount of heat is added to equal masses of A and B which are at the same initial temperature. After the heat has been absorbed, the temperature of A is greater than the temperature of B. Therefore the specific heat of A is

    a) greater than B   b) less than B   c) the same as B

    d) a negative value greater than B   e) the negative of B

17. If 25 J are required to change the temperature of 5.0 g of substance A by 2.0 K. What is the specific heat of substance A?

a) 250 J/ g· K   b) 63 J/ g· K   c) 10.J/ g· K

d) 2.5 J/ g· K   e) 0.40 J/ g· K

18. When 80. J of energy is absorbed by 0.50 mol of water, how much does the temperature rise? The specific heat of water is 4.184 J/ g· K.

a) 2.12 K   b) 38.2 K   c) 0.417 K   d)172 K   e) 3.22 K

19. When 75.4 J of energy is absorbed by 0.25 mol of CCl_4, what is the change in temperature? The specific heat of CCl_4 is 0.861 J/ g· K.

a) –17.8 K   b) –21.9 K   c) +2.3 K   d) +9.1 K   e) +44.6 K

20. When 0.25 mol of gold is cooled from 50.0°C to 20.0°C, what is the energy transferred? The specific heat of gold is 0.128 J/ g· K.

a) +0.96 J   b) +189 J   c) + 756 J   d) –189 J   e) –756 J

21. The specific heats of three elements are given. Rank the elements in order of their *molar* heat capacities.

Metal	Specific Heat (J/ g·K)
copper	0.385
magnesium	1.02
lead	0.129

a) Mg < Cu < Pb   b) Mg < Pb < Cu   c) Cu < Pb < Mg

d) Cu < Mg < Pb   e) Pb < Cu < <Mg

22. Consider 10.0 g Al, 10.0 g Cu and 10.0 g ethanol (C_2H_5OH). Which requires more energy for a 5.0°C temperature rise than does 1.0 gram of water?

Substance	Specific Heat (J/ g· K)
Ethanol (C_2H_5OH)	2.46
Water	4.184
Copper	0.385
Aluminum	1.02

a) 10.0 g Al

b) 10.0 g Cu

c) 10.0 g C_2H_5OH

d) both 10.0 g Al and 10.0 g C_2H_5OH

e) 10.0 g Al, 10.0 g Cu and 10.0 g C_2H_5OH

23. How many grams of lead will absorb the same amount of energy as 15.0 g Ag when each metals are heated from 20.0°C to 35.0 °C?

Substance	Specific Heat (J/ g· K)
Lead	0.129
Silver	0.237

a) 6.50 g   b) 27.5 g   c) 97.9 g   d) 53.3 g   e) 225 g

24. How much energy is required to change the temperature of 2.00 g of aluminum from 20.0°C to 25.0°C? The specific heat of aluminum is 0.902 J / g· K .

a) 2.26 J   b) 9.02 J   c) 0.361 J   d) 0.0902 J   e) 7.62 J

25. When 15.0 grams of an alloy is heated from 20.0°C to 40.0°C it absorbs 727 joules of energy . The specific heat of the alloy is

a) 2.42 J/g· K   b) 0.218 J/g· K   c) – 2.42 J/g· K

d) – 0.218 J/g· K   e) 0.206 J/g· K

26. When 17.0 grams of an alloy is cooled from 35.0°C to 10.0°C it releases 862 joules of energy . The specific heat of the alloy is

a) 0.493 J/g· K   b) 0.690J/g· K   c) 2.03 J/g· K

d) 3.66 J/g· K   e) 50.7 J/g· K

27. If 15.0 g water at 28.0°C is added to 125.0 g water at 20.0°C, what is the final temperature of the resulting mixture?

    a) 20.9 °C   b) 22.6 °C   c) 23.1 °C   d) 24.0 °C   e) 27.3 °C

28. If 12.0 g water at 35.0°C is added to 150.0 g water at 22.0°C, what is the final temperature of the resulting mixture?

    a) 22.1 °C   b) 23.0 °C   c) 25.2 °C   d) 28.5 °C   e) 31.7 °C

29. When 115 grams of water at 22.0°C is mixed with an unknown mass of water at a temperature of 58.0°C, the final temperature of the resulting mixture is 45.0°C. What was the mass of the second sample of water?

    a) 142 g   b) 187 g   c) 203 g   d) 265 g   e) 289 g

30. When 325 grams of water at 21.0°C is mixed with an unknown mass of water at a temperature of 45.0°C, the final temperature of the resulting mixture is 36.0°C. What was the mass of the second sample of water?

    a) 7.24 g   b) 226 g   c) 542 g   d) 2266 g   e) 874 g

31. When 150.0 grams of water at a temperature of 23.0°C is mixed with 45.0 grams of water an unknown temperature, the final temperature of the resulting mixture is 45.0°C. What was the temperature of the second sample of water.

    a) 53.5 °C   b) 58.5°C   c) 64.3°C   d) 68.9°C   e) 72.4 °C

32. A piece of metal rod weighing 3.20 grams is heated to 100.°C. It is dropped into 50.0 grams water in a calorimeter at 22.5°C. When no further change is observed, the temperature of the water and the metal rod is 26.5°C. What is the metal and its specific heat?

    a) Li , 3.56 J / g· K   b) Na , 1.23 J / g· K   c) Pb, 0.129 J / g· K
    d) Hg, 0.138 J / g· K   e) Cu , 0.385 J / g· K

33. A hot piece of aluminum weighing 0.500 grams at 350.°C is dropped into 500. gram of water at 22.0°C. what is the final temperature of the water. The specific heat of aluminum is 0.902 J/g· K and the specific heat of water is 4.184 J /g· K

    a) 85.0°C   b) 71.5 °C   c) 53.0 °C   d) 34.0 °C   e) 26.0 °C

34. Calculate the amount of heat needed to change 25.0 g ice at $- 15.0°C$ to steam at $100°C$.
    (Some constants for H_2O: Heat of fusion = 333 J/g; Heat of vaporization = 2260 J/g;
    Specific heats: Ice= 2.1 J / g· K, Water = 4.2 J / g· K, Steam 2.0 J / g· K)
    a) 75 kJ   b) 65 kJ   c) 48 kJ   d) 26 kJ   e) 11 kJ

35. Calculate the amount of heat needed to change 45.0 g ice at $-25.0°C$ to steam at $250°C$.
    (Some constants for H_2O: Heat of fusion = 333 J/g; Heat of vaporization = 2260 J/g;
    Specific heats: Ice= 2.1 J / g· K, Water = 4.2 J / g· K, Steam 2.0 J / g· K)
    a) 14.7 kJ   b) 33.7 kJ   c) 116 kJ   d) 150 kJ   e) 175 kJ

36. What are the signs of q and w for the exothermic process at a pressure of 1 atm and a
    temperature of 373 K for the following reaction?
    $$H_2O(g) \rightarrow H_2O(l)$$
    a) q is positive, w is negative   b) q is negative, w is negative
    c) q is positive, w is positive   d) q is negative, w is positive
    e) q is negative, w is zero

37. For a particular process q = 30 kJ and w = −25 kJ. What conclusion may be drawn for the
    process?
    a) ΔE = 55 kJ
    b) ΔE = −55 kJ
    c) the transfer of heat energy is from the system to the surroundings
    d) the transfer of work energy is from the surroundings to the system
    e) the system does work on the surroundings

38. What is ΔE for a system which has the following two steps.
    Step 1: The system absorbs 60 J of heat while 40 J of work are performed on it.
    Step 2: The system releases 30 J of heat while doing 70 J of work.
    a) 110 J   b) 100 J   c) 90 J   d) 30   e) zero

39. Consider the thermal energy transfer during a chemical process. When heat is transferred to
    the system, the process is said to be __?__ and the sign of q is __?__.
    a) exothermic, positive   b) exothermic, negative
    c) endothermic, positive   d) endothermic, positive
    e) enthalpic, negative

40. For the general reaction $2A + B_2 \rightarrow 2AB$, ΔH is $+50.0$ kJ. We can conclude that
    a) the reaction is endothermic
    b) the surroundings absorb energy
    c) the standard enthalpy of formation of AB is 50.0 kJ
    d) the bond energy of each A–B bond is 50.0 kJ
    e) the molecule AB contains less energy than A or B_2

41. When two solutions are mixed, the container "feels hot." Thus,
    a) the reaction is endothermic.
    b) the reaction is exothermic.
    c) the energy of the universe is increased.
    d) the energy of both the system and the surroundings is decreased.
    e) the energy of the system is increased

42. The equation for the standard enthalpy of formation of N_2O_3 is
    a) $N_2O(g) + O_2(g) \rightarrow N_2O_3(g)$ b) $N_2O_5(g) \rightarrow N_2O_3(g) + O_2(g)$
    c) $NO(g) + NO_2(g) \rightarrow N_2O_3(g)$ d) $N_2(g) + \frac{3}{2} O_2(g) \rightarrow N_2O_3(g)$
    e) $2NO(g) + 1/2\ O_2(g) \rightarrow N_2O_3(g)$

43. The equation for the standard enthalpy of formation of hydrazine, N_2H_4, is
    a) $2 N_2H_4(g) \rightarrow 2 NH_3(g) + H_2(g)$
    b) $2 NH_3(g) + H_2(g) \rightarrow N_2H_4(g)$
    c) $N_2(g) + 2 H_2O(g) \rightarrow N_2H_4(g) + O_2(g)$
    d) $N_2(g) + 2 H_2(g) \rightarrow N_2H_4(g)$
    e) $2 NO_2(g) + 6 H_2(g) \rightarrow N_2H_4(g) + 4 H_2O(g)$

44. The equation for the standard enthalpy of formation for magnesium nitrate $Mg(NO_3)_2$ (s) is
    a) $Mg(s) + 2 NO_3(g) \rightarrow Mg(NO_3)_2(s)$
    b) $Mg(s) + N_2(g) + 3 O_2(g) \rightarrow Mg(NO_3)_2(s)$
    c) $MgO_2(s) + 2 NO_2 \rightarrow Mg(NO_3)_2(s)$
    d) $2 MgO(s) + 2 N_2 + 5 O_2 \rightarrow 2 Mg(NO_3)_2(s)$
    e) $Mg_3N_2(s) + 6 NO_2 + 3 O_2 \rightarrow 3 Mg(NO_3)_2(s)$

45. Which of the following would have an enthalpy of formation value (ΔH_f) of zero?

    a) H_2O (g)   b) O (g)   c) H_2O (l)   d) O_2 (g)   e) H(g)

46. Which of the following equations represents an enthalpy change at 25°C and 1 atm that is equal to ΔH°_f?

    a) CO_2 (g) + H_2 (g) $\rightarrow$ HCOOH (l)

    b) CO (g) + H_2O (l) $\rightarrow$ HCOOH (l)

    c) 2 C (s) + 2 H_2 (g) + 2 O_2 (g) $\rightarrow$ 2 HCOOH (l)

    d) H_2O (l) + C(s) + 1/2 O_2 $\rightarrow$ HCOOH (l)

    e) C (s)   + H_2 (g) + O_2 (g) $\rightarrow$ HCOOH (l)

47. Which equation represents the standard enthalpy of formation for acrylonitrile, C_3H_3N?

    a) 3 C(graphite) + 3/2 H_2(g) + 1/2 N_2(g) $\rightarrow$ C_3H_3N(g)

    b) 3 C(graphite) + NH_3(g) $\rightarrow$ C_3H_3N(g)

    c) 3 CH_4(g) + NH_3(g) $\rightarrow$ C_3H_3N(g) + 6 H_2(g)

    d) 3 CO_2(g) + N_2 (g) + 3 H_2O (g) $\rightarrow$ C_3H_3N(g) + NH_3 + 9/2 O_2(g)

    e) 3 CO_2(g) + N_2 (g) + 3 H_2 (g) $\rightarrow$ C_3H_3N(g) + NH_3 + 3 O_2(g)

48. Which equation represents the standard enthalpy of formation for ethanol, C_2H_5OH?

    a) 2 CH_4(g) + H_2O(g) $\rightarrow$ C_2H_5OH(g)

    b) 2 CO_2(g) + 3 H_2O(g) $\rightarrow$ C_2H_5OH(g) + 3 O_2(g)

    c) CH_3OH(g) + CH_4 (g) $\rightarrow$ C_2H_5OH(g) + H_2 (g)

    d) 2 C(graphite) + 3 H_2(g) + 1/2 O_2(g) $\rightarrow$ C_2H_5OH(g)

    e) 2 C(graphite) + 3 H_2O(g) $\rightarrow$ C_2H_5OH(g) + O_2(g)

49. Given the heats of the following reactions:

    2 ClF (g) + O_2(g) $\rightarrow$ Cl_2O (g) + F_2O (g) ΔH = 167.4 kJ

    2 ClF_3(g) + 2 O_2(g) $\rightarrow$ Cl_2O (g) + 3 F_2O (g) ΔH = 341.4 kJ

    2 F_2(g) + O_2(g) $\rightarrow$ 2 F_2O(g) ΔH = – 43.4 kJ

    Calculate the heat of the reaction of ethylene with F_2 according to the equation:

    ClF(g) + F_2(g) $\rightarrow$ ClF_3(g)

    a) –108.7 kJ   b) –130.2 kJ   c) –217.5 kJ   d) +217.5kJ   e) +130.2 kJ

50. Calculate the heat of vaporization for titanium (IV) chloride

$$TiCl_4(l) \rightarrow TiCl_4(g)$$

using the following enthalpies of reaction

Ti (s) + 2 Cl_2(g) $\rightarrow$ $TiCl_4$ (l) ΔH^o =−804.2 kJ

$TiCl_4$ (g) $\rightarrow$ 2 Cl_2 (g) + Ti (s) ΔH^o = 763.2kJ

a) +41 kJ   b) −1567 kJ   c) 1165 kJ   d) −41 kJ   e) −783.7 kJ

51. Given the three step process:

C_3H_6(g) + H_2(g) $\rightarrow$ C_3H_8(g)

C_3H_8(g) + 5 O_2(g) $\longrightarrow$ 3CO_2(g) + 4 H_2O(l)

H_2O(l) $\rightarrow$ H_2(g) + 1/2O_2(g)

What is the equation for the overall reaction?

a) C_3H_6(g) + H_2O(l) + 5O_2(g) $\rightarrow$ 3 CO_2(g) + H_2(g)

b) 2 C_3H_6(g) + H_2(g) + 9/2O_2(g) $\rightarrow$ 3 CO_2(g) + 2H_2O(l)

c) C_3H_6(g) + 9/2O_2(g) $\rightarrow$ 3CO_2(g) + 2H_2O(l)

d) C_3H_6(g) + 2 H_2(g) + 5O_2(g) $\rightarrow$ 3CO_2(g) + 2H_2O(l)

e) C_3H_6(g) + 9/2 O_2(g) $\rightarrow$ 3CO_2(g) + 3H_2O(l)

52. Calculate the enthalpy of reaction for the process

A + 3 B $\rightarrow$ 2 C

using the following equations and data:

A + B $\rightarrow$ 2 E ΔH^o = −6 kJ /mol

C $\rightarrow$ B + E ΔH^o = + 2 kJ /mol

a) −12 kJ   b) -10 kJ   c) -8 kJ   d) + 4 kJ   e) +2 kJ

53. Calculate the enthalpy of reaction for the process

D + F $\rightarrow$ G + M

using the following equations and data:

G + C $\rightarrow$ A + B ΔH^o = +277

C + F $\rightarrow$ A ΔH^o = +303

D $\rightarrow$ B + M ΔH^o = -158

a) -132 kJ   b) +422 kJ   c) + 132 kJ   d) -184 kJ   e) -422 kJ

54. Calculate the enthalpy of combustion of C_3H_6 based on the equation
$$C_3H_6(g) + 9/2O_2(g) \rightarrow 3CO_2(g) + 3H_2O$$
    using the following data:

    $3C(s) + 3H_2(g) \rightarrow C_3H_6(g)$ $\Delta H^o = 53.3$ kJ

    $C(s) + O_2(g) \rightarrow CO_2(g)$ $\Delta H^o = -394$ kJ

    $H_2(g) + 1/2O_2(g) \rightarrow H_2O(l)$ $\Delta H^o = -286$ kJ

    a) -733 kJ   b) -626 kJ   c) -1517 kJ   d) 2090 kJ   e) 1304 kJ

55. Calculate the enthalpy for the decomposition of calcium carbonate according to the equation
$$CaCO_3(s) \rightarrow CaO(s) + CO_2(g)$$
    using the following

    $Ca(s) + 1/2O_2(g) \rightarrow CaO(s)$ $\Delta H^o = -635$ kJ

    $C(s) + O_2 \rightarrow CO_2(g)$ $\Delta H^o = -394$ kJ

    $Ca(s) + C(s) + 3/2O_2(g) \rightarrow CaCO_3(s)$ $\Delta H^o = -1207$ kJ

    a) 1447 kJ   b) 178 kJ   c) -1447 kJ   d) 966 kJ   e) -178 kJ

56. Calculate the enthalpy of reaction for the process
$$NO_2(g) + CO(g) \rightarrow CO_2(g) + NO(g)$$
    using the standard enthalpies of formation:

    $NO_2 = 34$ kJ/mol;   CO$= -111$kJ/mol; $CO_2 = -394$ kJ/mol;   NO $= 90$kJ/mol

    a) 339 kJ   b) 381 kJ   c) -227 kJ   d)-339 kJ   e)227 kJ

57. Calculate the standard enthalpy of reaction for the process
$$3\,NO \rightarrow N_2O + NO_2$$
    using the standard enthalpies of formation:

    NO$= 90$ kJ /mol; $N_2O= 82.1$kJ /mol; $NO_2= 34$kJ /mol

    a) 26.1 kJ   b) -153.9 kJ   c) -26.1 kJ   d) 206 kJ   e) 386 kJ

58. Calculate the standard enthalpy of formation of sulfuric acid from the following information.
$$Cu(s) + 2H_2SO_4(aq) \rightarrow CuSO_4(aq) + 2H_2O(l) + SO_2(aq) \qquad \Delta H^o_{rxn} = 178 \text{ kJ}$$
    $\Delta H_f^o(CuSO_4) = -771$ kJ/mol; $\Delta H_f^o(H_2O)= -286$ kJ/mol; $\Delta H_f^o(SO_2)= -297$kJ/mol

    a) -1532 kJ/mol   b) -766 kJ/mol   c) -1818 kJ/mol
    d) -909 kJ/mol   e) 1640 kJ/mol

59. Using the following information

$$C(s) + 2 Cl_2(g) \rightarrow CCl_4(l) \quad \Delta H^o = -135.4 \text{ kJ}$$

$$H_2(g) + Cl_2(g) \rightarrow 2 HCl(g) \quad \Delta H^o = -184.6 \text{ kJ}$$

$$2 H_2(g) + C(s) \rightarrow CH_4(g) \quad \Delta H^o = -74.8 \text{ kJ}$$

calculate the standard enthalpy of reaction for the process:

$$CH_4(g) + 4 Cl_2(g) \rightarrow CCl_4(l) + 4 HCl(g) \quad \Delta H^o_{rxn} = ?$$

a) 152.9 kJ   b) 302.1 kJ   c) 394.4 kJ   d) –429.8 kJ   e) 579.4 kJ

60. Given the following reactions

$$Fe_2O_3(s) + 3CO(g) \rightarrow 2 Fe(s) + 3CO_2(g) \quad \Delta H^o = -26.8 \text{ kJ}$$

$$FeO(s) + CO(g) \rightarrow Fe(s) + CO_2(g) \quad \Delta H^o = -16.5 \text{ kJ}$$

Calculate the enthalpy of the reaction for the process

$$Fe_2O_3(s) + 3CO(g) \rightarrow 2 Fe(s) + 3CO_2(g) \quad \Delta H^o_{rxn} = ?$$

a) +6.2 kJ   b) –6.2 kJ   c) –43.3 kJ   d) +43.3 kJ   e) 35.5 kJ

Chapter 6 : Answers to Multiple Choice

11. b	21. d	31. b
12. e	22. d	32. a
13. a	23. b	33. e
14. a	24. b	34. a
15. a	25. a	35. d
16. b	26. c	36. d
17. d	27. a	37. d
18. a	28. b	38. e
19. c	29. c	39. c
20. d	30. c	40. a

41. b	51. e
42. d	52. b
43. d	53. a
44. b	54. d
45. d	55. b
46. e	56. c
47. a	57. b
48. d	58. d
49. a	59. d
50. a	60. a

Chapter 7
Atomic Structure

Optional Reference

Equations: $E = h\upsilon$ $\upsilon\lambda = c$ $E = -Rhc/n^2$

$1/\lambda = R(1/n_{final}^2 - 1/n_{initial}^2)$ $\lambda = h/m\upsilon$

Avogadro's Number = 6.022×10^{23}
Planck's Constant = 6.626×10^{-34} J · s

R = 1.097×10^7 m^{-1}
Speed of Light = 2.998×10^8 m / s

Section A: Free Response

1. Compare the energy of one mole of yellow photons with a frequency of 5.33×10^{14} Hz to one mole of microwaves with a wavelength of 0.122 meters. Indicate which one has the greater energy and by how much.

2. Suppose a chemist in another planet in another galaxy performed the Millikan oil drop experiment . Recall that in the experiment electrons are absorbed onto oil droplets and the charge on the droplets can be observed. Some droplets contain one electron, some two electrons, some three, etc. The chemist observed the experimental charges in the table below.   Explain how this data can be to calculate the charge of the electron in tujohs (the unit for that planet).

Experiment	Charge Observed
#1	-3.55×10^{-15} tujohs
#2	-1.42×10^{-15} tujohs
#3	-4.26×10^{-15} tujohs
#4	-2.13×10^{-15} tujohs

3. Diagram the possible emission   n=4   _____
   transitions which can occur   n=3   _____
   for the atomic energy levels
   indicated for the H atom   n=2   _____

                                          n=1   _____

   Which transition emits photons with the highest frequency?

   Discuss the information concerning the atom that is gained by studying line spectra.

4. Match the names Bohr, Millikan, Planck, Rutherford, and Thomson to the discoveries listed.
    a) the atom has a dense positively charged nucleus
    b) emission spectra of elements are caused by electrons moving from high energy levels
        to low energy levels.
    c) radiant energy consists of packed called photons
    d) the charge on the electron was measured as -1.60×10^{-10} coulombs
    e) the charge/mass ratio for 20 different metals in the cathode ray tube was the same
        which suggested that electrons are present in all kinds of matter.

5. Einstein suggested that light waves may behave like particles. deBroglie suggested the converse of this idea. Discuss deBroglie's rationale and the experiment related to its acceptance. Use the deBroglie relationship to calculate the wavelength associated with an electron which has a velocity of 2.5×10^7 m/s. The mass of an electron is 9.109×10^{-31} kg.

6. The threshold frequency for the photoelectric effect of a particular metal is 5.0×10^{14} Hz. In three separate experiments, the metal is illuminated by light with a frequency of 1.0×10^{14} Hz, 5.1×10^{14} Hz and 7.0×10^{14} Hz. Compare and contrast the results of these three experiments.

7. No visible lines in the hydrogen spectrum correspond to an electron moving from the fifth shell (n=5) to the third shell (n=3). Predict whether the lines would lie in the ultraviolet or the infrared region and explain your logic. Calculate the wavelength of light that is emitted.

8. Classify each of the following sets of quantum numbers as valid or invalid. For those which are invalid, explain.
    a) $n = 3$, $\ell = 3$, $m_\ell = +1$
    b) $n = 3$, $\ell = 1$, $m_\ell = 0$
    c) $n = 3$, $\ell = 0$, $m_\ell = -1$
    d) $n = 4$, $\ell = 3$, $m_\ell = -2$
    e) $n = 4$, $\ell = 2$, $m_\ell = +2$

9. Explain what is being described by the quantum numbers n, ℓ, and m_ℓ.

10. a) For n=3, there are _____ types of orbitals which are labeled _____.
                    how many?

    b) For a 5d orbital, the allowable values of m_ℓ are _____.

    c) For n=4 there are _____ orbitals altogether _____.
               how many?   (number and label)

Key Concepts for Free Response:

1. The yellow photons are more energetic than the microwaves by a factor of 2.17×10^5.

2. By comparing the numbers and recognizing that each droplet must have a whole number of electrons, we calculate the lowest common factor as the charge on the electron, -7.3×10^{-16} tujohs.

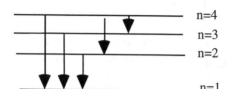

3. The transition with highest frequency also has the highest energy which is n=4 to n=1. By studying line spectra chemists can determine the difference in energy levels. For example, the line spectra have allowed us to see that as the value of n increases, the difference between successive energy levels is less. We rationalize this based on the decreasing effect of the nuclear pull on the electrons. We also see that many transitions are possible but the light emitted for a certain transition is constant.

4. a) Rutherford   b) Bohr   c) Planck   d) Millikan   e) Thomson

5. deBroglie suggested that particles may show wavelike properties according to the equation $\lambda = h/m\upsilon$. He noted that this was particularly important for very small particles (protons, neutrons, and electrons). Soon after his suggestion, scientists found that a beam of electrons is diffracted on a thin sheet of metal. This was behavior typical for waves and now was exhibited by electrons. Scientists then began to accept the wave/particle duality for matter. The wavelength associated with the electron is 2.9×10^{-11} meters.

6. No electrons will be emitted due to the photoelectric effect when 1.0×10^{14} Hz illuminates the metal. It has less too little energy to do anything. However, if the incident light is 5.1×10^{14} Hz, electrons will be emitted from the surface of the metal and travel to the anode and a current will flow. When the frequency is increased to 7.0×10^{14} Hz.

7. The lines should lie in the infrared region because a transition from n=5 to n=3 is much lower in energy than transitions from various levels to n=1. In the visible region, the final state is n=2. If the final energy state is n=3 or higher, the energy emitted is in the infrared region. From calculations the wavelength is 1.28×10^{-6} m.

8. The quantum number n describes the energy of the atomic orbital. It also is a measure of the distance from the nucleus. The quantum number ℓ indicates the shape of the orbital such as spherical, dumbbell, or cloverleaf. The orientation on the axes is described by the quantum number m_ℓ. Since some types of orbitals have several orientations, m_ℓ indicates a specific one of the available orbitals.

9. a) Invalid because ℓ cannot be 3 if n=3. There are no 3f orbitals.
   b) Valid   c) Invalid because m_ℓ cannot be −1 if ℓ=0. The only allowed value
   of m_ℓ is 0.   d) Valid.   e) Valid

10. a) For n=3, there are __3__ types of orbitals which are labeled __3s, 3p, 3d__.
                          how many?
    b) For a 5d orbital, the allowable values of m_ℓ are ___+1, 0, −1___.
    c) For n=4 there are ___16___ orbitals altogether _one 4s, three 4p, five 4d, and seven 4f_
                          how many?   (number and label)

Section B: Multiple Choice

11. Assume that the nucleus of an atom has a radius of 1.00 cm. What is the most reasonable value for the radius of the entire atom?
    a) 5.00 cm   b) 5.00 μm   c) 500 nm   d) 500 mm   e) 1.00 x 10^5 cm

12. Which of the following produces radiation of the highest frequency?
    a) x-rays   b) AM radio   c) FM radio   d) microwave oven   e) radar

13. Which of the following types of radiation has the lowest energy?
    a) gamma   b) visible   c) ultraviolet
    d) infrared   e) radio

14. What is the frequency of yellow light having a wavelength of 562 nanometers?
    a) 5.33 x 10^{14} s^{-1} b) 5.33x 10^5 s^{-1} c) 1.87 x $10^{-6}s^{-1}$
    d) 1.87 x 10^{-15} s^{-1} e) 1.18x $10^{-27}s^{-1}$

15. What is the frequency of ultraviolet radiation having a wavelength of 46.3 nanometers?
    a) 1.54 x 10^{-16} s^{-1} b) 6.47 x 10^6s^{-1} c) 1.54 x 10^{-9} s^{-1}
    d) 6.47 x 10^{15} s^{-1} e) 1.18 x $10^{-7}s^{-1}$

16. If a radio station has a frequency of 90.3 megahertz (MHz), what is the wavelength of the station in cm? (1 MHz = 1.00 x 10^6 cycles/second)
    a) 271 cm   b) 332 cm   c) 0.369 cm
    d) 110 cm   e) 0.254 cm

17. What is the frequency in megahertz of a radio station that is broadcasting at a wavelength of 284.0 cm? (1 MHz = 1.00 x 10^6 cycles/second)
    a) 85.1 MHz   b) 102.1 MHz   c) 99.4 MHz
    d) 95.6 MHz   e) 105.5 MHz

18. Planck suggested that all energy gained or lost by an atom must be some integral multiple of a minimum amount of energy called a(n)
    a) electron   b) spectrum   c) magnetic moment
    d) quantum   e) orbital

19. In the photoelectric effect, no electrons are emitted from the surface of a silver foil when the frequency of the incident light is less than 1.15×10^{15} Hz. What is the minimum energy necessary to eject an electron from the silver? (1 Hz =1 cycles/ second = 1 s^{-1})

a) 7.62×10^{-19} J   b) 3.44×10^{23} J   c) 1.74×10^{48} J

d) 1.26×10^{-22} J   e) 6.63×10^{-34} J

20. Which of the following transitions in the hydrogen atom results in the emission of light of the longest wavelength?

a) n=1 to n = 2   b) n=3 to n = 1   c) n = 2 to n = 1

d) n = 4 to n =3   e) n=1 to n=4

21. Which of the following transitions in the hydrogen atom results in the emission of light of the shortest wavelength?

a) n = 4 to n =3   b) n=1 to n = 2   c) n=3 to n = 1

d) n = 2 to n = 1   e) n=4 to n=1

22. If the frequency of an x-ray is 5.4×10^{18} Hz, what is the energy of one quantum of this radiation?

a) 3.6×10^{-15} J   b) 1.6×10^{-27} J   c) 1.2×10^{-52} J

d) 2.7×10^{-18} J   e) 7.4×10^{-29} J

23. If the frequency of a microwave is 3.8×10^{10} Hz, what is the energy of one quantum of this radiation?

a) 7.9×10^{-3} J   b) 1.1×10^{-19} J   c) 2.5×10^{-23} J

d) 4.1×10^{-47} J   e) 6.2×10^{-42} J

24. If wavelength of ultraviolet light is 105 nanometers, what is the energy of one quantum of this radiation?

a) 3.2×10^{-41} J   b) 1.89×10^{-18} J   c) 1.1×10^{-23} J

d) 8.7×10^{-29} J   e) 2.1×10^{-27} J

25. If the wavelength of blue light is 412 nanometers, what is the energy of one quantum of this radiation?
    a) 4.82×10^{-19} J   b) 7.31×10^{-14} J   c) 5.62×10^{-23} J
    d) 2.91×10^{-15} J   e) 8.36×10^{-29} J

26. What is the energy of a mole of photons of orange light with a wavelength of 585 nanometers?
    a) 1.61×10^{-27} J   b) 1.20×10^{-52} J   c) 2.78×10^{-18} J
    d) 2.04×10^{5} J /mol   e) 7.41×10^{-29} J

27. What is the energy of a mole of photons of infrared radiation of wavelength 1.72×10^{-3} cm?
    a) 6.95×10^{3} J/mol   b) 1.90×10^{5} J /mol   c) 1.04×10^{37} J /mol
    d) 2.83×10^{11} J /mol   e) 1.15×10^{20} J /mol

28. How many planar nodes are in the three orbitals represented?

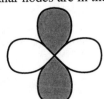

    a) 1, 2, 0 respectively   b) 2, 1, 0 respectively   c) 1, 2, 1 respectively
    d) 1, 2, 3 respectively   d) 2, 3, 1 respectively

29. Of the orbitals pictured represented, which one has the lowest value for the quantum number ℓ ?

    a)   b)   c)   d)   e)

30. What type of orbital is designated $n= 3$, $\ell=2$, $m_\ell = 0$
     a) 2s   b) 3s   c) 3p   d) 3d   e) 4d

31. What type of orbital is designated $n= 4$, $\ell=3$, $m_\ell = -1$
     a) 3p   b) 3d   c) 4p   d) 4d   e) 4f

32. What type of orbital is designated $n= 2$, $\ell=1$, $m_\ell = -1$
     a) 2p   b) 3p   c) 2s   d) 3s   e) 4p

33. What type of orbital (if any) is designated $n= 5$, $\ell=2$, $m_\ell = -2$
     a) 5p   b) 5p   c) 4d
     d) 5d   e) no orbital is identified

34. What type of orbital (if any) is designated $n= 3$, $\ell=1$, $m_\ell = -2$
     a) 3s   b) 3p   c) 2p
     d) 3d   e) no orbital is identified

35. Of the possible wave patterns for an electron, the pattern with only one planar node is described by which symbol?
     a) n   b) p   c) f   d) d   e) s

36. Of the possible wave patterns for an electron, the pattern with two planar nodes is described by which symbol?
     a) n   b) p   c) f   d) d   e) s

37. Which of the following orbitals would have three spherical nodes?
     a) 4s   b) 2p   c) 3p   d) 5f   e) 4d

38. What is the maximum number of orbitals having $\ell=1$ for a given electron shell greater than n=1?
     a) zero   b) one   c) three   d) five   e) nine

39. What is the maximum number of orbitals having $\ell=3$ for a given electron shell greater than n=3?
     a) zero   b) three   c) five   d) seven   e) nine

40. When $\ell=4$, what set of orbitals is designated?
   a) f   b) p   c) s   d) d   e) g

41. When $\ell=3$, what set of orbitals is designated ?
   a) g   b) p   c) f   d) d   e) s

42. Of the orbitals 1p, 2p, 3f, 4f, 5d, and 5g, quantum theory predicts which *cannot* exist?
   a) 1p, 3d, 4f and 5g   b) 1p and 3f   c) 1p, 3f and 5g
   d) 5g   e) 3f and 5f

43. Of the orbitals 2d, 3d, 3f, 4f, 5g, and 5h, quantum theory predicts which *cannot* exist?
   a) 2d, 3f and 5h   b) 2d, 5g, and 5h   c) 5h
   d) 5g and 5h   e) 2d, and 5g

44. What is the maximum number of orbitals that can be identified by the
    quantum numbers   n= 4, ℓ= 3, m$_\ell$ = −2 ?
   a) 0   b) 1   c) 3   d) 5   e) 7

45. What is the maximum number of orbitals that can be identified by the
    quantum numbers   n= 3, ℓ= 3, m$_\ell$ = −2 ?
   a) 0   b) 1   c) 3   d) 5   e) 7

46. Which of the following sets of quantum numbers is not allowed?
   a)   n= 3, ℓ= 3, m$_\ell$ = +1
   b)   n= 3, ℓ= 1, m$_\ell$ = 0
   c)   n= 3, ℓ= 0, m$_\ell$ = 0
   d)   n= 4, ℓ= 3, m$_\ell$ = −2
   e)   n= 4, ℓ= 2, m$_\ell$ = +2

47. Which of the following sets of quantum numbers is not allowed?
    a)   n= 1, ℓ= 0, m_ℓ = 0
    b)   n= 2, ℓ= 0, m_ℓ = 0
    c)   n= 2, ℓ= 2, m_ℓ = +1
    d)   n= 3, ℓ= 1, m_ℓ = 0
    e)   n= 3, ℓ= 1, m_ℓ = +1

48. Which of the following sets of quantum numbers is *not* allowed?
    a)   n= 2, ℓ= 1, m_ℓ = +1
    b)   n= 3, ℓ= 0, m_ℓ = −1
    c)   n= 4, ℓ= 2, m_ℓ = +1
    d)   n= 5, ℓ= 2, m_ℓ = 0
    e)   n=1, ℓ= 0, m_ℓ = 0

49. Which of the following sets of quantum numbers is not allowed?
    a)   n= 3, ℓ= 1, m_ℓ = +1
    b)   n= 3, ℓ= 0, m_ℓ = 0
    c)   n= 4, ℓ= 2, m_ℓ = +2
    d)   n= 4, ℓ= 1, m_ℓ = 0
    e)   n= 4, ℓ= 2, m_ℓ = +3

50. When n=2, which of the following is a possible value for ℓ?
    a) −2   b) 0   c) +2   d) 4   e) 8

51. When n=4, which of the following is a possible value for ℓ?
    a) −4   b) 2   c) 4   d) 8   e)16

52. When ℓ= 2, the possible values of m_ℓ are
    a) 0   b) 0, 1, 2   c) +1, 0, −1
    d) +2, +1, 0, -1, -2   e) +3. +2. +1, 0, −1, −2, −3

53. According to the Heisenberg's uncertainty principle, if one attempts simultaneously to measure the position and momentum of an electron, the more exactly the position is measured, the greater will be the _____ in the momentum measurement.
    a) probability   b) uncertainty   c) certainty
    d) polarity   e) energy

54. The quantum number ℓ represents the
    a) number of valence electrons   b) number of orbitals
    c) shape of the orbital   d) orientation of the orbital
    e) momentum of the electron

55. The quantum number m_ℓ represents the
    a) number of valence electrons   b) number of orbitals
    c) shape of the orbital   d) orientation of the orbital
    e) momentum of the electron

56. The number of orbitals in a 3f subshell is
    a) one   b) three   c) four   d) five   e) seven

57. The number of orbitals in a 4d subshell is
    a) one   b) four   c) five   d) eight   e) sixteen

58. According to the Bohr atomic theory, when an electron move from one energy level to another further from the nucleus
    a) energy is absorbed   b) energy is emitted   c) light is emitted
    d) photons are discharged   e) no change in energy is observed

59. Calculate the wavelength of light emitted when an electron changes from a state of principal quantum number 3 (n=3) to a state of principal quantum number 1 (n=1) in the H atom.
    a) 3.44×10^{-9}m   b) 7.24×10^{-4}m   c) 1.02×10^{-7}m
    d) 1.57×10^{-5}m   e) 2.75×10^{-37}m

60. In the photoelectric effect, no electrons are emitted from the surface of a silver foil when the frequency of the incident light is less than 1.15×10^{15} Hz. What is the minimum energy necessary to eject an electron from the silver?   (1 Hz =1 cycles/second = $1\ s^{-1}$)
    a) 6.63×10^{-34} J   b) 3.44×10^{23} J   c) 1.74×10^{48}J
    d) 1.26×10^{-22} J   e) 7.62×10^{-19} J

Chapter 7 : Answers to Multiple Choice

11. e	21. e	31. e
12. a	22. a	32. a
13. e	23. c	33. d
14. a	24. b	34. e
15. d	25. a	35. b
16. b	26. d	36. d
17. e	27. a	37. a
18. d	28. a	38. c
19. a	29. c	39. d
20. d	30. d	40. e

41. c	51. b
42. b	52. d
43. a	53. b
44. b	54. c
45. a	55. d
46. a	56. e
47. c	57. c
48. b	58. a
49. e	59. c
50. b	60. e

Chapter 8
Atomic and Electron Configurations and Chemical Periodicity

Section A: Free Response

1. Discuss the energy difference (if any) between the 2s and 2p orbitals of the hydrogen atom as compared to the energy difference between the 2s and 2p orbitals of the lithium atom.

2. Show the electron configuration of each of the following using the spectroscopic and noble gas notations: a) Mn   b) Sb^{3+} c) S^{2-}

3. Discuss the basis for the similar chemistry of the elements barium and calcium. Describe two atomic properties, based on their electronic configurations, in which these elements differ.

4. Explain why the first ionization energy of an element generally increases across a period of the Periodic Table.

5. Which element, Na or Mg, has the lower 2nd ionization energy? Explain fully.

6. Explain how the nuclear charge affects the sizes of positive and negative ions as compared to the parent atoms.

7. What is the maximum number of electrons that can be identified with each of the following sets of quantum numbers? If there are not any, write "none."

	n	ℓ	m_ℓ	m_s	Max number of electrons
a)	4				_____
b)	3	0	0		_____
c)	2	1	-1	1/2	_____
d)	4	4			_____

8. Of the chemical formulas V_2SO_4, VCl_5, VO_5, and VCO_3, which are consistent with the most likely ions of vanadium. For those which are not likely to exist, explain.

9. The ion Fe^{3+} is more paramagnetic than the ion Fe^{2+}. Explain how the measurement of paramagnetism allows us to determine the electron configurations of these ions.

10. The electron configuration of silver does not follow the usual "$n + \ell$" rule used for filling atomic orbitals. The experimental basis for this unusual pattern lies in the fact that silver chloride has the chemical formula AgCl and compound $AgCl_2$ does not exist. In light of this information show the electron configuration of silver and rationalize the pattern.

Key Concepts for Free Response:

1. The 2s and 2p orbitals have the same energy in the hydrogen atom but in the lithium atom the energy of the 2s is lower than the energy of the 2p orbital. The inner electrons of the lithium atom shield the electron in the 2s orbital from the full charge of the nucleus. Thus the effective nuclear charge is greater for an electron in the 2s orbital of lithium so it is held more tightly by the nuclear than an electron in the 2p orbital.

2. Mn $[Ar]4s^23d^5$ Sb^{3+} $[Kr]5s^24d^{10}5p^0$ S^{2-} $[Ne]3s^23p^6$

3. The similar chemistry is based on similar electron configurations with outer electrons in the ns^2 arrangement. Two electrons are easily lost to achieve a stable noble gas configuration. Thus both elements will react similarly with acids. Two atomic properties in which these elements differ are ionization energy and atomic radius. Barium has a lower ionization energy than calcium because the electrons are being lost from a higher level of n. They are held less tightly by the nucleus. The atomic radius of barium is larger than calcium because the electrons occupy a higher level of n.

4. Moving across the Periodic Table, the nuclear charge increases while the electrons are being added to the same level of n. This means that the electrons are held more tightly by the stronger nuclear charge and thus the ionization energy increases. It is harder to remove an electron with each successive element on a given level of n.

5. Magnesium has the lower 2nd ionization energy because it is easier to remove the 2nd electron for Mg which would be in the 3s orbital. The removal of 2 electrons to form Mg^{2+} is a stable electron configuration. On the other hand, for Na, removing the first electron is a low energy process but removing the second electron is a high energy event. The second electron would have to be from the full $2s^22p^6$ noble gas arrangement and would be quite high.

6. A positive ion contains more protons in the nucleus than electron surrounding it. Thus the size of the ion is much smaller than the atom because the nucleus can hold the electrons very tightly. On the other hand a negative ion has more electrons than protons in the nucleus. This decreases the pull of the nucleus on the electrons and the ions is substantially larger than the parent atom.

7. a) 32   b) 2   c) 1   d) none

8. Vanadium atoms have the electron configuration $[Ar]$ $4s^23d^3$ which indicates that V^{2+} and V^{5+} are the most likely ions of vanadium. Consistent with this are the compounds VCl_5 and VCO_3. The compound V_2SO_4 would have V^+ which is unlikely and VO_5 would have V^{10+} which is unlikely.

9. Iron atoms have the electron configuration $[Ar]$ $4s^23d^6$. If electrons are removed from the 4s orbital first and then the most energetic electron from the 3d, the ion Fe^{2+} and four unpaired electrons and Fe^{3+} has five unpaired electrons. This is consistent with the measurement of paramagnetism. However, if the electrons were removed from the 3d before the 4s, the ion Fe^{2+} would have 4 unpaired electrons and Fe^{3+} would have 3 unpaired electron. The experimental evidence does NOT support this electron arrangement.

10. Silver atoms must have the electron configuration $[Ar]$ $5s^14d^{10}$ rather than the predicted $[Ar]$ $5s^24d^9$ because the ion Ag^+ is present in AgCl and apparently Ag^{2+} does not form. Since the energy difference between levels 5s and 4d is small, the preferred arrangement of the 11 valence electrons of silver is $[Ar]$ $5s^14d^{10}$ in which the 4d electrons are all paired. This configuration is consistent with the experimental observations.

Section B: Multiple Choice

11. Of the orbitals depicted, the type most likely to be occupied by the 16th electron in the sulfur atom is

    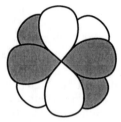

a)   b)   c)   d)

12. How many electrons can be described by the set of quantum numbers
    $n = 3$, $\ell = 1$, $m_\ell = -1$, $m_s = -1/2$?
    a) 18   b) 12   c) 1   d) 0   e) 6

13. How many electrons can be described by the set of quantum numbers
    $n = 4$, $\ell = 2$, $m_\ell = 0$, $m_s = +1/2$?
    a) 10   b) 5   c) 4   d) 1   e) 0

14. How many electrons can be described by the set of quantum numbers
    $n = 3$, $\ell = 3$, $m_\ell = -1$, $m_s = -1/2$?
    a) 18   b) 6   c) 2   d) 1   e) 0

15. How many electrons can be present in the n=5 shell?
    a) 5   b) 10   c) 25   d) 50   e) 54

16. What is the maximum number of orbitals that can be identified by the
    quantum numbers $n = 4$, $\ell = 3$, $m_\ell = -2$?
    a) 0   b) 1   c) 3   d) 5   e) 7

17. What is the maximum number of orbitals that can be identified by the
    quantum numbers $n = 2$, $\ell = 0$, $m_\ell = -2$?
    a) 0   b) 1   c) 2   d) 4   e) 8

18. Which of the following elements is paramagnetic?

   a) P   b) Mg   c) Zn   d) Ar   e) Ba

19. Which one of the following ions is paramagnetic?

   a) Zn^{2+} b) Ca^{2+} c) Ga^{3+} d) Ga^+ e) Fe^{3+}

20. Which one of the following ions is paramagnetic?

   a) F^- b) O^{2-} c) V^{2+} d) Sn^{2+} e) Ba^{2+}

21. Which of the following elements has three unpaired electrons in its a 3+ ion?

   a) aluminum   b) iron   c) chromium

   d) scandium   e) arsenic

22. Rank Na, Mg, Ca, and Zn in order of increasing 1st ionization energy.

   a) Na < Mg < Ca < Zn   b) Na < Ca < Mg < Zn   c) Zn < Mg < Ca < Na

   d) Ca < Mg < Zn < Na   e) Ca < Na < Mg < Zn

23. Rank C, O, Br, and Ne in order of increasing 1st ionization energy.

   a) C < Br < O < Ne   b) Ne < Br < O < C   c) C < O< Br < Ne

   d) Ne < O < C < Br   e) C < O < Ne < Br

24. Rank Na, Mg< and K in order of increasing 2nd ionization energy.

   a) Mg < Na < K   b) K < Mg < Na   c) Na < K < Mg

   d) K < Na < Mg   e) Mg < K < Na

25. Rank Ba, Ca, Na in order of increasing 2nd ionization energy.

   a) Ba < Ca < Na   b) Ba < Na < Ca   c) Ca < Ba < Na

   d) Na < Ca < Ba   e) Na < Ba < Ca

26. An element has the electron configuration $1s^2 2s^2 2p^6 3s^2 3p^6 4s^2 3d^{10}$
   If the element forms an ion, what is its charge?

   a) +1   b) +2   c) –2   d) –6   e) 0, no ion is possible

27. Which of the following is the correct electron configuration for the chromium (III) ion?

    a) [Ar] $4s^23d^4$ b) [Ar] $4s^03d^4$ c) [Ar] $4s^23d^2$

    d) [Ar] $4s^23d^6$ e) [Ar] $4s^03d^1$

28. Which of the following is the correct electron configuration for the nitride ion?

    a) $1s^22s^22p^3$ b) $1s^22s^22p^4$ c) [He]

    d) $1s^22s^22p^6$ e) $1s^22s^2$

29. What ion has the following electron configuration

    4s   3d

    a) Sc^{2+} b) V^{2+} c) Ca^{2+} d) Mn^{5+} e) Fe^{3+}

30. Which of the following elements has the most negative electron affinity?

    a) Cl   b) N   c) C   d) P   e) Na

31. A measure of the ability of an atom to acquire an electron to become negatively charged is called its

    a) ionization energy   b) polarizability   c) electron affinity

    d) electronegativity   e) electron density

32. Which of the following particles has the largest radius?

    a) Ne   b) F^- .   c) O^{2-} d) Na^+ e) N^{3-}

33. Which of the following particles has the lowest 1st ionization energy?

    a) F   b) O   c) Na   d) Mg   e) Ne

34. If the particles, F, F^- and O^{2-} are listed in an order of *increasing* diameter, the order is

    a) F, F^-, O^{2-} b) O^{2-}, F^-, F   c) F^-, F, O^{2-}

    d) O^{2-}, F, F^- e) F^-, O^{2-}, F

35. Which of the following particles would be predicted to be paramagnetic?

    a) Na   b) Ne   c) Mg   d) O^{2-} e) F^-

36. Which of the following particles would be most paramagnetic?
    a) P   b) Ga   c) Br   d) Cl^- e) Na^+

37. Which of the following has the smallest radius?
    a) Al   b) P   c) Sr   d) Ga   e) Na

38. Which of the following atoms has the largest number of valence electrons?
    a) Al   b) P   c) Sr   d) Ga   e) Ca

39. Which of the following atoms has the smallest first ionization energy?
    a) Al   b) P   c) Sr   d) Ga   e) Rb

40. Which of the following particles has the *lowest* 2nd ionization energy?
    a) F   b) O   c) Na   d) Mg   e) Li

41. For an atom other than hydrogen, which orbital has the lowest energy?
    a) 4f   b) 5d   c) 5f   d) 6s   e) 6d

42. The number of unpaired electrons in the selenium atom (Atomic Number 34) is
    a) 0   b) 2   c) 4   d) 6   e) 16

43. Which one of the following atoms has the largest atomic radius?
    a) Al   b) Ge   c) Ga   d)Si   e) P

44. If metallic character is characterized as the ability to lose electrons easily, the most
    metallic of the atoms Mg, Ba, Sn, and Pb is
    a) Ba   b) Mg   c) Pb   d) Sn   e)Zn

45. How many unpaired electrons are present in Fe^{+2}.
    a) 0   b) 2   c) 4   d) 5   e) 6

46. Of the particles K^+, Ca^{2+} , S^{-2-} , and Cl^-, which one, if any, has the largest in radius?
    a) K^+ b) Ca^{2+} c) S^{2-} d) Cl^- e) they are all the same

47. When arranged in order of increasing atomic number, the elements exhibit periodicity
for all of the following properties *except*
   a) 1st ionization energy.   b) atomic radii.   c) atomic masses.
   d) electron affinity   e) 2nd ionization energy

48. Which of the following sets of quantum numbers is not allowed?
   a)   n= 1, $\ell = 0$, $m_\ell = 0$, $m_s = -1/2$
   b)   n= 2, $\ell = 1$, $m_\ell = +1$, $m_s = +1/2$
   c)   n= 3, $\ell = 0$, $m_\ell = -1$, $m_s = +1/2$
   d)   n= 4, $\ell = 2$, $m_\ell = +1$, $m_s = -1/2$
   e)   n= 5, $\ell = 2$, $m_\ell = 0$, $m_s = -1/2$

49. Which of the following sets of quantum numbers is not allowed?
   a)   n= 1, $\ell = 0$, $m_\ell = 0$, $m_s = 0$
   b)   n= 2, $\ell = 1$, $m_\ell = +1$, $m_s = +1/2$
   c)   n= 2, $\ell = 0$, $m_\ell = 0$, $m_s = +1/2$
   d)   n= 3, $\ell = 2$, $m_\ell = +1$, $m_s = -1/2$
   e)   n= 3, $\ell = 2$, $m_\ell = 0$, $m_s = -1/2$

50. Which of the following sets of quantum numbers is not allowed?
   a)   n= 3, $\ell = 2$, $m_\ell = 0$, $m_s = -1/2$
   b)   n= 3, $\ell = 2$, $m_\ell = +2$, $m_s = +1/2$
   c)   n= 2, $\ell = 2$, $m_\ell = -1$, $m_s = +1/2$
   d)   n= 4, $\ell = 2$, $m_\ell = +1$, $m_s = -1/2$
   e)   n= 4, $\ell = 3$, $m_\ell = -3$, $m_s = -1/2$

51. Which of the following most probably cannot exist?
   a) $BaBr_2$ b) KBr_2 c) Br_2 d) OBr_2 e)$FeBr_2$

Chapter 8 : Answers to Multiple Choice:

11. a	21. c	31. c
12. c	22. b	32. e
13. d	23. a	33. c
14. e	24. e	34. a
15. e	25. a	35. a
16. b	26. b	36. a
17. a	27. b	37. b
18. a	28. d	38. b
19. e	29. b	39. e
20. c	30. a	40. d

41. d	51. b
42. b	
43. c	
44. a	
45. c	
46. c	
47. c	
48. c	
49. a	
50. c	

Chapter 9
Bonding and Molecular Structure: Fundamental Concepts

Section A: Free Response

1. Dinitrogen oxide, N_2O, is also called nitrous oxide or "laughing gas" and is sometimes used as an anesthetic. Circle the correct Lewis structures below.   Explain what is wrong with the other two structures.

   Proposed Lewis Structures:

   #1   #2   #3

2. Use the concept of formal charge with Lewis structures to explain why the sulfur atom in sulfuric acid (H_2SO_4) has an expanded octet rather than the octet.

3. Give an explanation for the observation that the molecular geometry of NH_3 is pyramidal rather than triangular planar.

4. Explain why CO_2 is a non-polar molecule while SO_2 is a polar molecule.

5. Draw the Lewis structures for IF_2^- and IF_2^+. Use your structures to discuss the electron pair geometry, the molecular geometry and the predicted F–I–F bond angle in each ion.

6. Consider the reaction below in which two bonds are broken and two bonds are formed. Explain why the enthalpy of reaction is not zero.
   $$H_2(g) + Cl_2(g) \text{-------}> 2\,HCl(g) \qquad \Delta H_{rxn} = -183 \text{ kJ}$$

7. Phosphorus forms two chlorides, PCl_3 and PCl_5. Nitrogen, in the same group as phosphorus, forms the chloride NCl_3, but not NCl_5. Explain the difference in behavior between these two elements.

8. Compare the nitrogen-nitrogen bond length in the molecules N_2H_4, N_2, and N_2O_4. Discuss reasons for a difference (if any).

9. In each of the following sets of compounds, pick the one with the highest melting point and explain your choice:

Set 1: NaF, NaCl, NaI   Set 2: MgO, $MgBr_2$, BaO

10. For each of the following compounds, draw the Lewis structure. Discuss the shape of the molecule based on VSEPR Theory and indicate the relevant bond angles. Use your structure to discuss the dipole moment of the molecule (if any).

ClF_2^+ I_3^- PCl_4^+

Key Concepts for Free Response:

1. Structure #1 is correct for N_2O. #2 is wrong because oxygen is more electronegative than N and must be in a terminal position in the skeletal structure. #3 has too many electrons.

2. The 32 electrons can be distributed without expanding the octet as in Structure #1. However, this structure gives a formal charge of 1– on each oxygen and a 2+ on sulfur. By forming double bonds between sulfur and oxygen as in Structure #2 both sulfur and oxygen have a formal charge of zero. The lower the formal charge the more likely the structure.

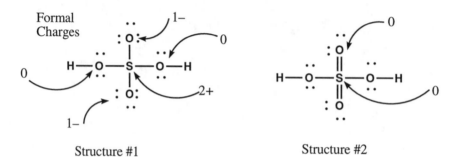

Structure #1   Structure #2

3. The nitrogen has an unshared pair of electrons so there are four structural pairs and a pyramidal molecular shape.

4. CO_2 has only 16 valence electrons as 8 pairs. The Lewis structure shows that the molecule is linear with a double bond between each carbon and oxygen. The bonds are polar but the molecule is non-polar since the oxygen atoms are pulling in opposite directions on the carbon. It has no dipole moment.
   SO_2 has 18 valence electrons so the Lewis structure dictates an unshared pair on the sulfur and a trigonal planar molecular geometry. This gives SO_2 an uneven distribution of electrons and hence a polarity. It has a dipole moment.

5. There are 5 structural pairs around the central I atom in IF_2^-. The structural pair geometry is triangular bipyramidal having three long pairs and two shared pairs. Since the long pairs occupy the equatorial space the F—I—F bond angle is 180°.

In contrast IF_2^+ has 4 structural pairs around the iodine atom. The geometry is tetrahedral with two lone pairs. The bond angles are 109.5°.

6.   Different bonds contain different energies. The energy required to break one H–H bond and one Cl–Cl bond is less than the energy released when two H–Cl bonds are formed. Therefore we see that the net reaction is exothermic.

7.   Phosphorus is in the 3rd Row of the Periodic Table and can have an expanded octet as in PCl_5.   The larger diameter of the n=3 shell makes it possible to place more "things" around it. On the other hand, nitrogen does not have the ability to expand its octet because the valence electrons are on the energy level n=2. Nitrogen cannot form NCl_5.

8.   The Lewis structures show a N=N in N_2H_4, N–N bond in N_2O_4 and N≡N in N_2. The longest bond is in N_2O_4, an intermediate length in N_2H_4 and the shortest bond is in N_2.

9.   Set 1: NaF because it has the strongest attraction between ions. The charge density on the fluoride ion is larger than for the other halides because of its smaller size.
      Set 2: MgO because it has the strongest attraction between ions. The 2+ and 2– ions fit neatly into a crystal and have a strong ionic bond.

10. ClF_2^+ is tetrahedral with bond angles about 109.5°. It does have a dipole with the fluorine end being more negative.

     I_3^- is linear with bond angles of 180°. It has three unshared pairs of electrons on the central iodine but it does have not have a dipole because it is very symmetrical.

     PCl_4^+ is tetrahedral with bond angles about 109.5°. The molecule does not have a dipole since all the Cl atoms are pulling in different directions on the phosphorus.

Section B: Multiple Choice

11. Which species has more than eight electrons around the central atom?
    a) BF_3 b) BF_4^- c) BrF_3 d) PF_3 e) OF_2

12. Which of the following compounds exhibits ionic bonding?
    a) CCl_4 b) $MgCl_2$ c) Cl_2 d) PCl_3 e) OF_2

13. Describe the shape and polarity of the molecule O=C=Se .
    a) It is linear and polar with the O end more positive than the Se end.
    b) It is linear and polar with the Se end more positive than the O end.
    c) It is angular with the carbon end more positive.
    d) It is linear and non-polar.
    e) It is angular and non-polar.

14. Which is the most polar bond?
    a) O—F   b) N—F   c) C—F   d) F—F   e) Cl—F

15. In drawing Lewis structures for the following compounds, which one would have oxygen as
     the central atom ?

    a) Na_2O b) NO_2 c) N_2O d) OF_2 e) SOCl

16. The Lewis structure represented is

$$:\overset{\cdot\cdot}{\underset{\cdot\cdot}{O}}—N\overset{\cdot\cdot}{=}\overset{\cdot\cdot}{\underset{\cdot\cdot}{O}}:$$

    a) NO_2 b) NO_2^- c) NO_2^+
    d) both NO_2^- and NO_2^+ e) NO_2, NO_2^- , and NO_2^+

17. The Lewis structure represented is

$$:\overset{\cdot\cdot}{\underset{\cdot\cdot}{O}}=N=\overset{\cdot\cdot}{\underset{\cdot\cdot}{O}}:$$

    a) NO_2 b) NO_2^- c) NO_2^+
    d) both NO_2^- and NO_2^+ e) NO_2, NO_2^- , and NO_2^+

18. Which of the following best describes the energy change accompanying the process of
    breaking bonds in a molecule? (Ignore any subsequent reaction that may occur.)
    a) Always endothermic.
    b) Always exothermic
    c) The net energy change is always zero.
    d) The change may be exothermic or endothermic depending on the physical state.
    e) The change may be exothermic or endothermic depending on the substances
        involved.

19. Which of the following elements is most likely to form compounds involving an
    expanded valence shell of electrons?
    a) Li   b) N   c) F   d) Ne   e) S

20. Which of the following element combinations is likely to produce ionic bonds in a
    compound?
    a) lithium and fluorine   b) boron and oxygen   c) nitrogen and oxygen
    d) phosphorus and sulfur   e) chlorine and bromine

21. Which of the following element combinations is likely to produce covalent bonds in a
    compound?
    a) potassium and fluorine   b) magnesium and oxygen   c) nitrogen and chlorine
    d) sodium and chlorine   e) sodium and fluorine

22. Which compound has the most ionic bond?
    a) LiCl   b) LiF   c) KF   d) KCl   e) NaCl

23. Which compound has the most ionic bond?
    a) $CaCl_2$ b) BrCl   c) CO_2 d) CCl_4 e) CO_2

24. Which of the following groups contains *no* ionic compounds?
    a) H_2O, MgO, NO_2 b) CO_2, SO_2, H_2S
    c) CCl_4, $CaCl_2$, HCl   d) Na_2S, SO_2, CS_2
    e) Mg_3N_2, NCl_3, HOCl

25. Which of the following compounds would have the highest melting point?
    a) LiF   b) LiCl   c) NaBr   d) CsF   e) NaCl

26. Which of the following compounds would have the lowest melting point?
    a) NaCl   b) LiF   c) LiCl   d) CsF   e) CsI

27. Which of the following groups of elements is arranged in order of increasing electronegativity?
    a) $K < Mg < O$ b) $Na < K < N$ c) $Ca < Cl < C$
    d) $Br < Li < C$ e) $Mg < Li < C$

28. Which of the following groups of elements is arranged in order of increasing electronegativity?
    a) $Si < Al < Br < Cl$ b) $Na < K < Ca < Ba$ c) $P < S < O < F$
    d) $K < Rb < Cs < F$ e) $N < P < S < Cl$

29. Which compound contains a carbon-oxygen bond with an order of 2?
    a) CO_2 b) CH_3OH c) $CH_3—O—CH_3$
    d) CO e) C_2H_5OH

30. Use the Lewis structure to determine the bond order of the nitrogen-oxygen bond in NO_2^+

$$\left[\ddot{:O}\!=\!=\!N\!=\!=\!\ddot{O:} \right]^+$$

    a) 0   b) 1   c) 2   d) 3   e) 4

31. Which statement is true regarding bond order, bond length, and bond energy.
    a) As the bond order increases, the bond length increases.
    b) As the bond order increases, the bond length decreases.
    c) As the bond order increases, the bond energy decreases.
    d) As the bond energy increases, the bond length increases.
    e) As the bond energy increases, the bond order decreases.

32. Which of the following diatomic molecules has the greatest bond strength?

    a) F_2 b) O_2 c) N_2 d) HF   e) HCl

33. For a molecule which has 3 resonance structures, a plausible bond order between the
    central atom and an electron negative atom is

    a) 0   b) 1   c) 2   d) 3/2   e) –1

34. What is the average carbon-oxygen bond order in the formate ion ?

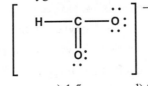

    a) 0   b) 1   c) 1.5   d) 2   e) 2.5

35. Given the bond dissociation energies below, calculate the standard molar enthalpy of
    formation of ClF_3.

$$Cl_2(g) + 3\ F_2(g) \ \text{-----} > \ 2\ ClF_3(g)$$

Bond	Dissociation Energy (kJ /mol)
Cl–Cl	243
F–F	159
Cl–F	255

    a) –45 kJ/mol   b) – 33 kJ/mol   c) -405 kJ/mol

    d) +210 kJ/mol   e) +147 kJ/mol

36. Given the bond dissociation energies below, calculate the standard molar enthalpy of
    formation of OCN.

$$C(s) + 1/2\ O_2(g) + 1/2\ N_2(g) \ \text{-----} > \ O=C=N(g)$$

Bond	Dissociation Energy (kJ /mol)
O=O	498
N≡N	946
C=O	745
C=N	615

    a) +84 kJ   b) +638 kJ   c) –226 kJ

    d) –638 kJ   e) –84 kJ

37. What is the formal charge on each atom in the following structure for the nitrite ion, NO_2^-?

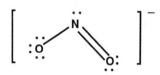

a) nitrogen is 2–, oxygen on the left is 1–, oxygen on the right is 0
b) nitrogen is 0, oxygen on the left is 0, oxygen on the right is 1–
c) nitrogen is 0, oxygen on the left is 1–, oxygen on the right is 0
d) nitrogen is 3–, oxygen on the left is 1–, oxygen on the right is -2
e) nitrogen is 1+, oxygen on the left is 2–, oxygen on the right is 1–

38. What is the formal charge on sulfur in the molecule, SO_3?

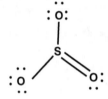

a) 0   b) 1+   c) 1–   d) 2+   e) 2–

39. What are the formal charges on sulfur and oxygen in the ion SO_3^{2-} ?
a) sulfur is 1+ and oxygen is 1–
b) sulfur is 6+ and oxygen is 2–
c) sulfur is   0 and oxygen is 2–
d) sulfur is 2+ and oxygen is 0
e) sulfur is 4+ and oxygen is 1–

40. What is the oxidation number of sulfur in H_2SO_4?
a) +8   b) +7   c) +6   d) +4   e) –2

41. What are the oxidation numbers of sulfur and oxygen in the molecule SO_3?
    a) sulfur is +1 and oxygen is –1
    b) sulfur is +6 and oxygen is -2
    c) sulfur is +6 and oxygen is –6
    d) sulfur is +3/2 and oxygen is –3
    e) sulfur is -2 and oxygen is -2

42. What is the oxidation number of chlorine in the compound ClF_3?
    a) -1   b) 0   c) +1   d) +3   e) +6

43. According to the VSEPR Theory, what number of structural electron pairs is normally expected to produce a tetrahedral structural-pair geometry?
    a) two   b) three   c) four   d) five   e) six

44. According to the VSEPR Theory, what number of structural electron pairs is normally expected to produce a triangular bipyramidal structural-pair geometry?
    a) two   b) three   c) four   d) five   e) six

45. According to the VSEPR Theory, what number of structural electron pairs is normally expected to produce a triangular planar structural-pair geometry?
    a) two   b) three   c) four   d) five   e) six

46. Based on the VSEPR Theory, what is the molecular shape of ClF_3?
    a) triangular planar   b) T-shaped   c) linear
    d) tetrahedral   e) square planar

47. Based on the VSEPR Theory, what is the molecular shape of SCl_2?
    a) triangular planar   b) T-shaped   c) linear
    d) tetrahedral   e) angular (bent)

48. Based on the VSEPR Theory, what is the molecular shape of CF_4?
    a) triangular planar   b) T-shaped   c) linear
    d) tetrahedral   e) angular (bent)

49. Based on the VSEPR Theory, what is the molecular shape of the ion ICl_4^-?

    a) triangular planar   b) angular (bent)   c) octahedral

    d) tetrahedral   e) square planar

50. Based on the VSEPR Theory, what is the molecular shape of NH_3?

    a) triangular planar   b) T-shaped   c) triangular-pyramidal

    d) tetrahedral   e) octahedral

51. If a molecule has five structural electron pairs and the molecular structure is linear, how many lone pairs are present in the molecule?

    a) one   b) two   c) three   d) four   e) zero

52. If a molecule has six structural electron pairs and the molecular structure is square pyramidal, how many lone pairs are present in the molecule?

    a) one   b) two   c) three   d) four   e) zero

53. If a molecule has three structural electron pairs and the molecular structure is triangular-planar, how many lone pairs are present in the molecule?

    a) one   b) two   c) three   d) four   e) zero

54. Two ions which have a similar shape are

    a) CO_3^{2-} and NO_3^- b) CO_3^{2-} and SO_3^{2-} c) PO_3^{3-} and CO_3^{2-}
    d) PO_3^{3-} and NO_3^- e) SO_4^{2-} and NO_3^-

55. Two ions which have a similar shape are

    a) OH^- and SO_3^{2-} b) SO_3^{2-} and CO_3^{2-} c) PO_3^{3-} and CO_3^{2-}
    d) PO_3^{3-} and NO_3^- e) SO_3^{2-} and PO_3^{3-}

56. How many unshared election pairs (lone pairs) are in a molecule of SO_2?

    a) 2   b) 6   c) 7   d) 9   e) 12

57. What is the approximate C—O—H bond angle in CH_3OH ?

    a)180°   b) 120°   c) 109.5°   d) 90°   e) 60°

58. What is the approximate H—C—C bond angle in $H_2C=CH_2$?
    a)180°   b) 120°   c) 109.5°   d) 90°   e) 60°

59. What are the approximate bond angles of 1, 2, and 3 respectively?

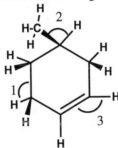

    a) 120°, 120°, 180°   b) 109.5°, 120°, 180° c) 109.5°, 120°, 120°
    d) 180°, 120°, 120°   e) 109.5°, 109.5°, 180°

60. What are the approximate bond angles of 1, 2, and 3 respectively?

    a) 120°, 120°, 180°   b) 109.5°, 109.5°, 109.5°   c) 109.5°, 120°, 120°
    d) 120°, 120°, 120°   e) 109.5°, 109.5°, 120°

61. Which of the following molecules has no dipole moment?
    a) BH_3 b) NO_2 c) SF_6 d) PCl_5 e) $CHCl_3$

Chapter 9 : Answers to Multiple Choice:

11. c	21. c	31. b
12. b	22. c	32. c
13. b	23. a	33. d
14. c	24. b	34. c
15. d	25. a	35. c
16. b	26. e	36. d
17. c	27. a	37. c
18. a	28. c	38. d
19. e	29. a	39. a
20. a	30. c	40. c

41. b	51. c	61. c
42. d	52. a	
43. c	53. e	
44. d	54. a	
45. b	55. e	
46. b	56. b	
47. e	57. c	
48. d	58. b	
49. e	59. b	
50. c	60. e	

Chapter 10 Bonding and Molecular Structure:
Orbital Hybridization, Molecular Orbitals, and Metallic Bonding

Section A: Free Response

1. Draw the Lewis structure of the molecule N_2F_2.
   Use your structure to complete the following:

   The hybridization of nitrogen is _____.
   <small>atomic type</small>

   Therefore, the nitrogen has _____ hybridized _____ orbital(s) and
   <small>how many</small>   <small>atomic type</small>

   _____ unhybridized _____ orbital(s).
   <small>how many</small>   <small>atomic type</small>

2. What is the hybridization of the central atom in each particle below. Determine the bond
   angles about the central atom.

particle	hybridization of central atom	bond angles
FClO		
SF_5^-		

3. The two molecules in each set below have the same empirical formula. Are they
   identical? Use principles of bonding to explain fully.

Set 1	Set 2
H H Cl Cl │ │ │ │ Cl-C-C- Cl H-C-C-H │ │ │ │ H H H H	Cl Cl Cl H \C=C/ \C=C/ / \ / \ H H H Cl

   Set 1   Set 2

4. Discuss the hybridization of the carbon atoms in allene, $H_2C=C=CH_2$. Describe the σ
    and π bonding the in the molecule and the resulting geometric relationship of the
    hydrogen atoms with respect to each other.

Reference for Molecular Orbital section:

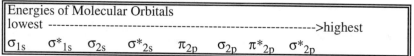

| Energies of Molecular Orbitals |
| lowest -->highest |
| σ_{1s} σ^*_{1s} σ_{2s} σ^*_{2s} π_{2p} σ_{2p} π^*_{2p} σ^*_{2p} |

5. Consider the four species N_2, N_2^+, O_2, and O_2^+. Based on their MO diagrams, which has the highest bond energy and which has the lowest bond energy. List the species (if any) which are paramagnetic. Calculate the bond order for each particle.

6. Use the molecular orbital model to describe the differences between the bond energies of N_2 and F_2.

7. Draw the molecular orbital diagram including the electrons for the ion O_2^{2-}. Use your diagram to determine the bond order and the magnetic character of the ion.

8. Explain why the Lewis structure is inadequate for the molecule O_2.

9. Use the molecular orbital theory to explain why the bonding in metals is unique. How does the observation that metals are good conductors of electricity support the theory?

10. Explain how elements are combined to form a semiconductor. What elements can be used with silicon?

Key Concepts for Free Response

1. The hybridization of nitrogen is sp^2. Therefore, the nitrogen has 3 hybridized sp^2 orbitals and 1 unhybridized p orbital.

2. FClO sp^3 hybridization   bond angles are 109.5^o
   SF_5^- sp^3d^2 hybridization   bond angles are 90° or 180°

3. The two molecules in Set 1 are identical because there is free rotation about all the s bonds which attached the atoms. In Set 2 the molecules are different because the pi bond restricts rotation about the two atoms and structural isomers are formed. The molecule on the left in Set 2 is the cis compound and would have a dipole toward the Cl end. The molecule on the right is the trans compound and would have no dipole.

4. The carbon atoms on the ends are sp^2 hybridized with one unhybridized p orbital . The center carbon atom is sp hybridized with two unhybridized p orbitals perpendicular to each other. The overlap of the p orbitals forms a π bonding each case. However, the π bonds are perpendicular to each other so the hydrogen atoms are in perpendicular planes also.

5. Highest Bond Energy N_2 Lowest Bond Energy: O_2
   paramagnetic particles $N_2^+ O_2$ O_2^+
   Bond Orders: $N_2 = 3$ $N_2^+ = 2.5$ $O_2 = 2$ $O_2^+ = 2.5$

6. The bond order of N_2 is 3 and the bond order for F_2 is 1. Therefore N_2 has a higher bond strength and a higher bond energy than F_2. It would require more energy to break the bond in N_2 than in F_2.

7. The valence electrons are the ones that count so $2(6) + 2 = 14$ electrons in the following MO's. The bond order is $(8-6)/2 = 1$ All the electrons are paired so it is diamagnetic.

   σ^*_{2p} _____
   π^*_{2p} ↑↓   ↑↓
   σ_{2p} ↑↓
   π_{2p} ↑↓   ↑↓
   σ^*_{2s} ↑↓
   σ_{2s} ↑↓

8. The Lewis structure has all electrons paired and a complete octet on the oxygen. However, it does not allow us to explain the experimental observation that O_2 is attracted to the poles of a strong magnet (paramagnetic). This property indicates unpaired electrons. The Molecular Orbital theory give a diagram that would be paramagnetic.

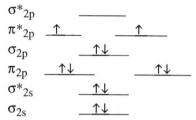

9. Large numbers of metallic atoms are grouped together not in a normal kind of chemical bond but rather as an association of many atoms linked by delocalized bonds. The MO's formed between atoms are lower in energy than the atomic orbitals and are partly filled. This allows metals to be good conductors of electricity because it is easy for electrons to flow through the groups of atoms. Metals also have unoccupied MO's that are nearly the same energy as the occupied MO's so it is easy for an electron to be promoted to a higher level during conduction. If the MO's were filled with electrons or if a large energy difference between the filled and unfilled MO's existed, no conductivity would be possible.

10. Semiconductors have a small band gap meaning a small energy difference between the highest occupied MO's and other empty MO's. When electrons are promoted to a higher level a molecular "switch" is present that can be used for many practical purposes. Pure silicon is a semiconductor. Its properties can be enhanced by the addition of elements from the adjacent rows in the periodic table. If an element with more electrons than Si (Group 5A, elements such as P or As) is added, the donor level of the band gap is lowered and the n-type semiconductor is formed. If an element with fewer electrons than Si (Group 3A, elements such as Al or Ga) is added, the acceptor level of the band gap is raised which gives a reduced band gap and a p-type semiconductor.

Section B: Multiple Choice

11. How many sigma (σ)bonds are in the following molecule?

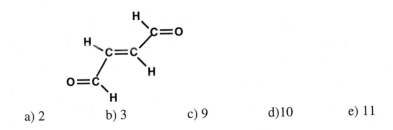

    a) 3   b) 7   c) 8   d) 9   e) 12

12. How many pi (π) bonds are in the following molecule? $CH_3C{\equiv}C{-}C{\equiv}N$

    a) 2   b) 4   c) 6   d) 7   e)10

13. How many sigma bonds (σ) are in the following molecule?

    H₂N
        \
         C=O
        /
    H₂N

    a) 3   b) 4   c) 6   d) 7   e) 8

14. How many pi (π) bonds are in the following molecule?

            H
             \
              C=O
         H   /
          \ /
           C=C
          /   \
      O=C   H
          \
           H

    a) 2   b) 3   c) 9   d)10   e) 11

15. How many sigma(σ) and pi (π) electron pairs are in a carbon dioxide molecule?

    a) four σ and zero π   b) three σ and two π

    c) two σ and two π   d) two σ and four π

    e) one σ and three π

16. How many sigma(σ) and pi (π) electron pairs are in a nitrogen molecule, N_2.
    a) one σ and three π b) one σ and two π
    c) two σ and two π d) two σ and three π
    e) one σ and three π

17. If an atom uses an sp^2 hybrid orbital set, how many *unhybridized* p orbitals in the same energy level remain on the atom?
    a) 0 (zero)   b) 1   c) 2   d) 3   e) 4

18. If an atom uses an sp^3d hybrid orbital set, how many *unhybridized* d orbitals in the same energy level remain on the atom?
    a) 0 (zero)   b) 1   c) 3   d) 4   e) 5

19. A molecule, QX_3, which has the electron pairs around the central Q atom arranged in a tetrahedral fashion probably utilizes what hybridization of the Q atomic orbitals?
    a) sp   b) sp^2 c) sp^3 d) sp^3d e) sp^3d^2

20. All the S–F bonds in SF_5^- are identical. The hybridization of the sulfur atomic orbitals which most likely accounts for this identity is
    a) sp   b) sp^2 c) sp^3 d) sp^3d e) sp^3d^2

21. What is the hybridization of the carbon atom in CS_2 ?
    a) sp   b) sp^2 c) sp^3 d) sp^3d e) sp^3d^2

22. What is the hybridization of the sulfur atom in SCl_4 ?
    a) sp^3 b) sp^4 c) sp^3d d) sp^3d^2 e) sp^2d^2

23. What is the hybridization of the carbon atoms in CH_3OH and CO_2 respectively?
    a) sp^3, sp^3 b) sp^3, sp^2 c) sp^2, sp^2
    d) sp^2 , sp^3 e) sp^3, sp

24. What is the hybridization of the oxygen atoms in CH_3OH and CO_2 respectively?
    a) sp^3, sp^3 b) sp^3, sp^2 c) sp^2, sp^2
    d) sp^2, sp^2 e) sp^3, sp

25. What is the hybridization of the nitrogen atoms in NH_3 and NH_4^+ respectively?

    a) sp^3, sp^4 b) sp^3, sp^3 c) sp^2, sp^3

    d) sp^2, sp^2 e) sp^3, sp

26. What is the hybridization of the sulfur atom in SO_3?

    a) sp^2 b) sp^3 c) sp^4 d) sp^3d e) sp^3d^2

27. An atom in which hybridization state (if any) can form pi (π) bonds.

    a) sp^2 b) sp^3 c) sp^3d

    d) sp^3d^2 e) no hybridization is possible.

28. In the combustion of methane, CH_4, what change in hybridization (if any) occurs to the carbon atom?

    a) sp^2 to sp^3 b) sp^2 to sp^3 c) sp^2 to sp^3

    d) sp^3 to sp e) no change in hybridization occurs

29. In the addition of fluorine to xenon difluoride to form xenon tetrafluoride, what change in hybridization (if any) occurs to the xenon atom?

    a) sp^3d^2 to sp^3d b) sp^2 to sp^3d c) sp^3 to sp^3d^2

    d) sp^3d to sp^3d^2 e) no change in hybridization occurs

Optional Reference for Molecular Orbital Section:

Energies of Molecular Orbitals

lowest --->highest

σ_{1s} σ^*_{1s} σ_{2s} σ^*_{2s} π_{2p} σ_{2p} π^*_{2p} σ^*_{2p}

Energy level diagram

σ^*_{2p} ———

π^*_{2p} ———   ———

σ_{2p} ———

π_{2p} ———   ———

σ^*_{2s} ———

σ_{2s} ———

30. Consider the diatomic molecules of the second period Li_2, Be_2, B_2, and C_2. Which is(are) unlikely to exist?

a) Li_2 b) Li_2 and Be_2 c) Be_2

d) C_2 e) Be_2 and C_2

31. Consider the diatomic molecules of the second period Li_2, Be_2, C_2, and Ne_2. Which is(are) unlikely to exist?

a) Li_2 b) Li_2 and Be_2 c) C_2

d) Ne_2 e) Be_2 and Ne_2

32. Consider the diatomic molecules of the second period Li_2, Be_2, B_2, and C_2. The two molecules which have the same bond order are

a) Li_2 and Be_2 b) Li_2 and B_2 c) Li_2 and C_2

d) Be_2 and B_2 e) Be_2 and C_2

33. Consider the diatomic molecules of the second period C_2, N_2, O_2, and F_2. The two molecules which have the same bond order are

a) C_2 and N_2 b) N_2 and O_2 c) O_2 and F_2

d) N_2 and O_2 e) C_2 and O_2

34. Based on bond orders, list the species , B_2^+, B_2, and B_2^-, in the order of increasing bond length.

    a) B_2^+, B_2, B_2^- b) B_2, B_2^+, B_2^- c) B_2^-, B_2, B_2^+

    d) B_2^-, B_2^+, B_2 e) B_2^+, B_2^-, B_2

35. The species below having the longest bond (if any) is

    a) N_2^+ b) N_2 c) N_2^-

    d) N_2^{2-} e) they are all the same length

36. The species below having the shortest bond (if any) is

    a) N_2^+ b) N_2 c) N_2^-

    d) N_2^{2-} e) they are all the same length

37. The bond order of carbon monoxide (CO) is most probably

    a) 0   b) 1   c) 2   d) 3   e) 4

38. The following molecular orbital energy level diagram is appropriate for which one of the listed particles?

    σ^*_{2p} _____

    π^*_{2p} _____   _____

    σ_{2p} _____

    π_{2p} ⤒⤓   ⤒

    σ^*_{2s} ⤒⤓

    σ_{2s} ⤒⤓

    a) B_2^+ b) B_2^- c) N_2^+ d) N_2^- e) N_2

39. The following molecular orbital energy level diagram is appropriate for which one of the listed particles?

σ^*_{2p} _____

π^*_{2p} ↑_____   ↑_____

σ_{2p} ↑↓_____

π_{2p} ↑↓_____   ↑↓_____

σ^*_{2s} ↑↓_____

σ_{2s} ↑↓_____

   a) O_2^{2-} b) B_2 c) F_2^- d) N_2^{2-} e) N_2

40. The molecular orbital configuration of He_2^+ is
   a) $(\sigma_{1s})^1 (\sigma^*_{1s})^1$
   b) $(\sigma_{1s})^2 (\sigma^*_{1s})^2$
   b) $(\sigma_{1s})^2 (\sigma^*_{1s})^1$
   d) $(\sigma_{1s})^2 (\sigma^*_{1s})^2 (\sigma_{2s})^2$
   e) $(\sigma_{1s})^2 (\sigma^*_{1s})^2 (\sigma_{2s})^2 (\sigma^*_{2s})^2$

41. The molecular orbital configuration of O_2 is
   a) [core electrons] $(\sigma_{2s})^2 (\sigma^*_{2s})^2 (\pi_{2p})^4 (\sigma_{2p})^2 (\pi^*_{2p})^1 (\sigma^*_{2p})^1$
   b) [core electrons] $(\sigma_{2s})^2 (\sigma^*_{2s})^2 (\pi_{2p})^4 (\sigma_{2p})^2 (\pi^*_{2p})^2$
   c) [core electrons] $(\sigma_{2s})^2 (\sigma^*_{2s})^2 (\pi_{2p})^2 (\sigma_{2p})^2 (\pi^*_{2p})^2 (\sigma^*_{2p})^2$
   d) [core electrons] $(\sigma_{2s})^2 (\sigma^*_{2s})^2 (\pi_{2p})^2 (\sigma_{2p})^2 (\pi^*_{2p})^1 (\sigma^*_{2p})^1$
   e) [core electrons] $(\sigma_{2s})^2 (\sigma^*_{2s})^2 (\pi_{2p})^4 (\sigma_{2p})^2 (\pi^*_{2p})^2 (\sigma^*_{2p})^2$

42. The molecular orbital configuration of B_2 is
   a) [core electrons] $(\sigma_{2s})^2 (\sigma^*_{2s})^2 (\pi_{2p})^4 (\sigma_{2p})^2$
   b) [core electrons] $(\sigma_{2s})^2 (\sigma^*_{2s})^2 (\pi_{2p})^4 (\sigma_{2p})^1 (\pi^*_{2p})^1$
   c) [core electrons] $(\sigma_{2s})^2 (\sigma^*_{2s})^2 (\pi_{2p})^2 (\sigma_{2p})^1$
   d) [core electrons] $(\sigma_{2s})^2 (\sigma^*_{2s})^2 (\pi_{2p})^1$
   e) [core electrons] $(\sigma_{2s})^2 (\sigma^*_{2s})^2 (\pi_{2p})^2$

43. Of the molecules, B_2 , C_2 , N_2, O_2, and F_2 , which is (are) diamagnetic?
   a) B_2 and O_2 b) C_2 and O_2 c) N_2 and O_2
   d) N_2 and F_2 e) N_2 and C_2

44. Of the species, N_2, N_2^+, N_2^-, and N_2^{2-} , how many are paramagnetic ?

    a) four   b) three   c) two   d) one   e) zero (none)

45. Which of the following is a diamagnetic particle?

    a) He_2^+ b) N_2^- c) N_2 d) He_2^- e) O_2

46. In forming a metallic solid, how many orbitals are contributed by each lithium atom to a group of atoms?

    a) four, the 2s and three 2p orbitals

    b) three, the three 2p orbitals

    c) two, the 1s and 2s orbitals

    d) two, the 2s and one 2p orbital

    e) one, the 2s orbital

47. What name is given to the highest filled energy level in a metal at absolute zero?

    a) conduction band   b) Fermi level   c) extrinsic orbital

    d) band gap   e) antibonding orbital

48. Which substance has cations bonded together by mobile electrons?

    a) Ag(s)   b) S_8(s)   c) Br_2(l)   d) KBr(s)   e) $MgCl_2$(s)

49. In order to create an *n-type* semiconductor, a silicon crystal could be doped with

    a) Ga   b) Ge   c) As   d) He   e) none of these

50. In order to create an *p-type* semiconductor, a silicon crystal could be doped with

    a) Ga   b) Ge   c) As   d) He   e) none of these

51. How do the number of electrons in the valance band of a metal compare with an insulator?
    a) completely full for a metal; partially filled for an insulator
    b) completely full for a metal; completely full for an insulator
    c) partially filled for a metal; partially filled full for an insulator
    d) partially filled for a metal; completely full for an insulator
    e) completely filled for a metal; totally empty for an insulator

Chapter 10 : Answers to Multiple Choice

11. d	21. b	31. e
12. b	22. c	32. b
13. d	23. e	33. e
14. b	24. a	34. c
15. c	25. b	35. d
16. b	26. a	36. d
17. b	27. a	37. d
18. d	28. d	38. b
19. c	29. d	39. d
20. e	30. c	40. b

41. b	51. d
42. e	
43. a	
44. b	
45. c	
46. a	
47. b	
48. a	
49. c	
50. a	

Chapter 11
Bonding and Molecular Structure: Organic Chemistry

Section A: Free Response

1. What is the chemical formula of the principal hydrocarbon in gasoline? Write the balanced chemical equation for the combustion of this molecule?

2. Draw three structures for the formula $C_2H_2Br_2$. Name each molecule and discuss its polarity.

3. The bond lengths of several carbon-carbon bonds are given. Based on this information, which of the following bond lengths, 115 pm, 121 pm, 134 pm, 139 pm, 154 pm, or 169 pm is most reasonable for the carbon–carbon bonds in benzene? Explain your choice. Some Approximate bond lengths: C–C 154 pm; C=C 134 pm; C¿C 121 pm .

4. Match the active ingredient of each product to its chemical formula:
   products: gasoline   antifreeze   handcream   natural gas
   formulas: $HOCH_2CH_2OH$, CH_4
                  $HOCH_2CH(OH)CH_2OH$, C_8H_{18}

5. Give the systematic name for each of the following compounds. What is the hybridization of each carbon atom? What are the bond angles about each carbon atoms?

$$\underset{HCH}{\overset{O}{\overset{\|}{}}} \qquad \underset{HCOH}{\overset{O}{\overset{\|}{}}} \qquad CH_3OH \qquad HC{\equiv}CH$$

6. Write equations using some of the following molecules to illustrate the preparation of an ester and the preparation of an amide.

$$\underset{C{\cdot}O{\cdot}H}{\overset{O}{\overset{\|}{}}} \qquad CH_3OH \qquad NH_2CH_3 \qquad H_2SO_4$$

7. Discuss the difference between polyunsaturated and unsaturated fats with respect to their chemical structure, chemical properties, and health hazard.

8. Of the molecules aspirin, methanol, and ethylene, which one is produced in largest quantity by U.S. industries? Discuss some uses of the molecule.

9. Which of the following compounds could NOT be used in a condensation polymerization reaction with $HOCH_2CH_2OH$? Explain your choice.

10. Explain the difference between high density polyethylene and low density polyethylene.

Key Concepts for Free Response:

1. Octane is C_8H_{18}. The combustion reaction is
   $$C_8H_{18} + 25/2\ O_2 \longrightarrow 8\ CO_2 + 9\ H_2O$$

2. The three compounds are 1,2-dibromoethane which is polar, *cis*-1,2-dibromoethane which is polar, and *trans*-1,2-dibromoethane with is not polar. The *trans* compound is not polar because the electronegative atoms are symmetrically arranged about the double bond.

3. Benzene has two resonance structures so the carbon-carbon bond length is greater than the single bond but less than the double bond. A value of 139 pm is reasonable.

4. antifreeze $HOCH_2CH_2OH$; natural gas CH_4; glycerol $HOCH_2CH(OH)CH_2OH$; gasoline C_8H_{18}.

5. The compounds are methanal, sp^2 hybridization, 120° bond angles; methanoic acid, sp^2, 120° ; methanol, sp^3 , 109.5° ; ethyne, sp, 180°.

6. Equations to synthesize an ester and an amide are

7. Polyunsaturated fats contain many double bonds as part of the long chain of the ester. The rigid geometry about the double bond causes the molecules to be "bent". Therefore polyunsaturated fats have a lower melting point than their corresponding saturated fats since the "bent" molecules will not fit into a neatly packed solid structure. They are often oils rather than solids. The unsaturated fats are more reactive especially toward oxidation. This means that foods with unsaturated fats tend to spoil more easily. In general saturated fats are more harmful to your health than are unsaturated fats.

8. Ethylene is produced in largest quantity as a molecule that serves the producers of a large variety of plastics, particularly polyethylene. Polyethylene can be branched to form a low density plastic (LDPE) which is flexible and soft such as that used in plastic sandwich bags. If it is not branched is forms a high density plastic (HDPE) which is stronger and used in materials such as milk bottles.

9. Compound #2 is already a diester and could not undergo another condensation polymerization.

10. High density polyethylene has long unbranched chains which fit neatly into a lattice in the solid state. The melting point is high and the polymer has a tough strong property. Low-density polyethylene is branched, so the long chains do not fit into a crystalline pattern compactly. The melting point is low and the polymer is soft and flexible.

Section B: Multiple Choice

11. Which of the following formulas could be an alkane?
    a) C_2H_4 b) C_3H_7 c) C_4H_{10} d) C_5H_8 e) C_6H_6

12. Which of the following formulas could be a cycloalkane?
    a) C_2H_4 b) C_3H_7 c) C_4H_{10} d) C_5H_8 e) C_6H_6

13. Which of the following formulas could be an alkyne?
    a) C_2H_4 b) C_3H_7 c) C_2H_2 d) C_5H_8 e) C_6H_{14}

14. The chemical formula of acetylene is ?
    a) CH_4 b) C_2H_2 c) C_6H_6 d) C_2H_4 e) CH_3CO_2H

15. How many of the following compounds are unsaturated?

$$H_3C-\underset{\underset{CH_3}{|}}{\overset{\overset{CH_3}{|}}{C}}-CH_2\ CH_3 \qquad H_3C-\underset{}{\overset{\overset{CH_3}{|}}{C}}=CH-CH_3 \qquad CH_3C\equiv CH \qquad H_3C-CH_3$$

    a) zero   b) 1   c) 2   d) 3   e) 4

16. The formula for the following compound is

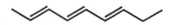

    a) C_9H_{14} b) C_9H_{12} c) C_7H_8 d) C_7H_{12} e) C_8H_{12}

17. The formula for the following compound is

a) C₆H₆   b) C₆H₁₀   c) C₆H₁₂   d) C₈H₁₄   e) C₈H₁₆

18. The formula for the following compound is

a) C₈H₆   b) C₆H₆   c) C₆H₁₀   d) C₈H₁₀   e) C₈H₁₂

19. How many of the following compounds are aromatic molecules?

a) zero   b) 1   c) 2   d) 3   e) 4

20. The common name of the following compound is

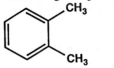

a) *o* -dimethylbenzene   b) *p* -dimethylbenzene   c) *m* -dimethylbenzene
d) 2,3-dimethylbenzene   e) toluene

21. The name of the following compound is

$$CH_2CH_2CH_3$$
$$H_3C-\underset{\underset{CH_3}{|}}{\overset{|}{C}}-CH_2\ CH_3$$

a) 2,2-diethylpropane   b) 2-methyldibutane   c) 2-methyloctane
d) 2-methyl-2-propylbutane   e) 3,3-dimethylhexane

22. The name of the following compound is

$$CH_3$$
$$H_3C-\underset{\underset{CH_3}{|}}{\overset{|}{C}}-CH_2\ CH_3$$

a) 2,2,3-trimethylpropane   b) 2,2-dimethylpropane   c) 2,2-dimethyl butane
d) 2,2-dimethylhexane   e) 3,3-dimethylhexane

23. Which of the following are polar molecules?

          #1   #2   #3

a) #1 only   b) #2 only   c) #3 only
d) both #1 and #2   e) both # 2 and #3

24. The balanced equation for the combustion 2-butene is
   a) $C_4H_6\ +\ 6\ O_2\ ---> 4\ CO_2 + 3\ H_2O$
   b) $C_4H_{10}\ +\ 4\ O_2\ ---> 4\ CO_2 + 5\ H_2$
   c) $C_4H_8\ +\ 4\ O_2\ ---> 4\ C + 4\ H_2O$
   d) $C_4H_8\ +\ 6\ O_2\ ---> 4\ CO_2 + 4\ H_2O$
   e) $C_4H_{10}\ +\ 13/2\ O_2\ ---> 4\ CO_2$ and $5\ H_2O$

25. A ban on the production of which of the following molecules was imposed in January 1996?

    a) CCl_3F b) CCl_4 c) CF_4 d) $CHCl_3$ e) CF_2H_2

26. What is the product of the addition of Br_2 to $H_2C=CH_2$?

    a) 1,1-dibromoethylene   b) 1,1, dibromoethane

    c) 1,2-dibromoethylene   d) 1,2-dibromoethane

    e) 1,2-dibromocyclopropane

27. What is the product of the addition of H_2O to $H_2C=CH_2$?

    a) ethanol   b) methanol   c) 1-propanol

    d) 2-propyl alcohol   e) 1,2-ethandiol

28. Rubbing alcohol is which propanol?

    a) 1-propanol   b) 2-propanol   c) 3-propanol

    d) cyclpropanol   e) 1,2-propanediol

29. Which of the following alcohols is the poisonous "wood alcohol"?

    a) glycerol   b) ethanol   c) methanol

    d) 1-propanol   e) 2-propanol

30. Which of the following molecules is a secondary alcohol?

    a)   b)   c)   d)   e)

31. An isomer of dimethyl ether $CH_3–O–CH_3$ is

    a) $CH_3CO_2CH_3$ b) $CH_3CH_2–O–CH_2CH_4$ c) $HOCH_2CH_2OH$

    d) CH_3CO_2H e) CH_3CH_2OH

32. When an alcohol is dehydrated, the product could be an

    a) alkane   b) alkene   c) aldehyde   d) acid   e) ester

33. When an alcohol is oxidized, the product could be an

    a) alkane   b) alkene   c) aldehyde   d) alkyne   e) isomer

34. Which group of compounds includes an aldehyde, an acid, and an alcohol (in any order)?

a) HCO_2H, $CH_3CO_2CH_3$, CH_3CH_2OH

b) H_2CO, CH_3CH_2OH, $CH_3CO_2CH_3$

c) CH_3CO_2H, H_3COH, $CH_3CH_2OCH_3$

d) H_2CO, CH_3CO_2H, CH_3CH_2OH

e) H_2CO, CH_3CO_2H, CH_3CH_2OH

35. Lactic acid, found in milk, has the formula

a)   b)   c)   d)   e)

36. Which of the following compounds is an ester?

a)   b)   c)   d)   e)

37. When methanol is heated with benzoic acid, in the presence of acid, the product is

a)   b)   c)   d)   e)

38. The following molecule is classified as

a) an aromatic acid   b) a fatty acid   c) a soap

d) an aldehyde   e) a polyester

39. Which of the following molecules is a fat?

    a)  CO_2H

    b)  CO_2H

    c) 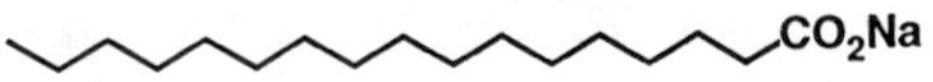 CO_2Na

    d) 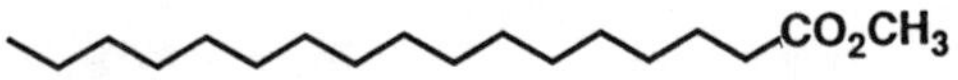 CO_2CH_3

    e)  H_2CO_2C
            HCO_2C
            H_2CO_2C

40. The monomer of polystyrene is

    a) $C_6H_5CH=CH_2$ b) $C_6H_5CH_3$ c) $C_6H_5CH_2CH_3$

    d) $H_2C=CH_2$ e) $H_2C=CHCH=CH_2$

41. Polyethylene

    a) contains equal numbers of cis and trans double bonds.

    b) contains no double bonds.

    c) reacts with methanol to form polyester.

    d) cannot form branched chains.

    e) is an example of a condensation polymer.

42. Low density polyethylene

    a) packs together easily with long linear molecules.

    b) has a branched structure.

    c) is used to make tough rigid materials such as milk cartons.

    d) cannot be recycled.

    e) contains long chains of alternating double bonds.

43. Vulcanized rubber contains long hydrocarbon chains linked with
    a) short chains of sulfur.
    b) short chains of ethylene.
    c) long chains of silicon.
    d) long chains of unsaturated hydrocarbons.
    e) boric acid, $B(OH)_3$.

44. A copolymer of styrene and butadiene is used to make what common product?
    a) coffee cups   b) plastic wrap for foods   c) bubble gum
    d) paint   e) nylon

45. The condensation polymer formed by a diacid and a dialcohol is classified as a
    a) a composite   b) soap   c) a polyurethane
    d) a polyamide   e) a polyester

46. Nylon-6,6 can be produced from the condensation polymerization of a diacid with six carbon atoms and a diamine with six carbon atoms. Nylon would be classified as a
    a) polyamine   b) polyurethane   c) a polyalcohol

        d) a polyamide   e) a polyester

Chapter 11 : Answers to Multiple Choice

11. c	21. e	31. e
12. a	22. c	32. c
13. c	23. d	33. e
14. b	24. d	34. b
15. c	25. a	35. d
16. a	26. d	36. a
17. d	27. a	37. a
18. d	28. b	38. b
19. b	29. c	39. e
20. c	30. b	40. d

41. b

42. b

43. a

44. c

45. e

46. d

Chapter 12
Gases

Section A: Free Response

1. An impure sample contains aluminum and an unreactive material. A sample weighing 5.00 grams is analyzed by its reaction with hydrochloric acid according to the following equation. If 4.16 liters of hydrogen collected over water at 25.0°C and 0.845 atm is produced, what is the percentage of aluminum in the sample? The vapor pressure of water at 25.0°C is 23.8 torr.

$$2 \text{ Al (s)} + 6 \text{ HCl (aq)} \longrightarrow 3 \text{ H}_2 \text{ (g)} + 2 \text{ AlCl}_3 \text{ (aq)}$$

2. A sample of liquid nitrogen trichloride was heated in a 1.50 liter closed container until it decomposed completely to gaseous elements. The resulting mixture exerted a pressure of 744 torr at 75°C. a) What is the partial pressure of each gas in the container? b) What was the mass of the original sample?

$$2 \text{ NCl}_3 \text{ (l)} \longrightarrow \text{N}_2 \text{ (g)} + 3 \text{ Cl}_2 \text{ (g)}$$

3. The density of a gas composed only of carbon and fluorine is 5.61 g/L at 22°C and 748 torr. If the gas is 17.40% carbon, find the molecular formula for this gas.

4. The graph below represents the distribution of molecular speeds for Ne and another noble gas which are present together in a flask with a total pressure of 700 torr. The partial pressure of Ne is measured as 200 torr. Is the other gas helium or argon? Explain fully.

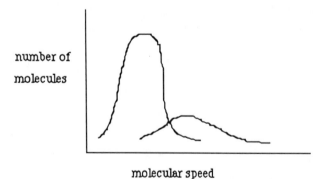

5. A compound contains 54.52 % carbon, 9.17% hydrogen, and 36.31% oxygen. A sample weighing 3.023 grams has a volume of 2.00 liters at 25.0°C and 0.420 atm. Calculate a) the empirical formula   b) the molar mass and c) the molecular formula.

6. A sample of 0.0500 mol of ammonia gas had a volume of 235 mL. The temperature and pressure were held constant and some ammonia was removed from the sample. Its new volume was 175 mL. How many grams of ammonia were removed?

7. Explain why it is necessary to use Kelvin rather than Celcius temperatures in gas law problems.

8. A sample of neon was collected over water and stored immediately in Gas Bulb X as pictured. How would the same apparatus appear if the same sample of neon at the same room conditions is collected directly (not collected over water) and stored in an identical gas bulb. Select one of the choices and explain fully.

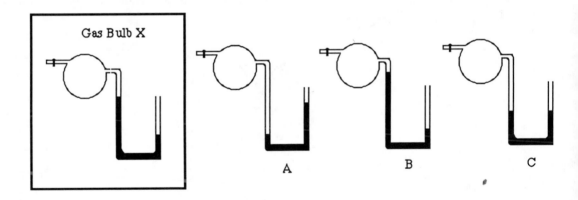

9. Consider three identical flasks filled with different gases.

Flask X: CO_2 at 250 torr and 10°C

Flask Y: N_2 at 760 torr and 10°C

Flask Z: H_2 at 100 torr and 10°C

a) In which flask (if any) will the molecules have the greatest average kinetic energy?

b) In which flask (if any) will the molecules collide with the wall with the greatest frequency?

c) In which flask (if any) will the molecules have the greatest root mean square velocity?

d) In which flask does the gas have the largest density?

10. Gases A and B react to form the gas C according to the equation

$$2 A(g) + B(g) \longrightarrow 3 C(g)$$

Consider the reaction of 6.0 mol Gas A in an 8.0 L vessel at 0.0°C and 1.0 mol Gas B in an 8.0 liter vessel at 0.0°C as diagrammed. The gases mix freely when the stopcock is opened. What is the limiting reagent? How many moles of each gas is present at the completion of the reaction? Calculate the *total* pressure in the apparatus at the completion of the reaction. Assume no change in temperature.

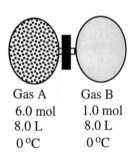

Gas A   Gas B
6.0 mol   1.0 mol
8.0 L   8.0 L
0 °C   0 °C

Key Concepts for Free Response

1. The sample is 48.1% Al.

2. a) Pressure of N_2 is 186 torr or 0.245 atm, pressure of Cl_2 is 558 torr or 0.734 atm
   b) 3.10 g NCl_3

3. Molecular Formula is C_2F_6

4. The partial pressure of Ne is 200 torr so the partial pressure of the other gas is 500 torr. This indicates that there are fewer moles of neon and neon is the graph with the lesser area. The other gas has a greater amount and a slower speed than Ne. It is Argon.

5. a) C_2H_4O b) 88.04 g/mol   c) $C_4H_8O_2$

6. 0.17 g of ammonia were removed

7. The Kelvin scale is based on a point of no molecular motion, absolute zero. Thus all the temperatures are compared to the motion of the particles. The Celcius scale is a *relative* scale with measurements being compared to the boiling and freezing points of water. The Kelvin scale is appropriate for gas law problems because the changes in volume and pressure for a gas are based on the motion of the particles.

8. The gas was collected wet, so it had water vapor contributing to the total pressure. Had the gas been collected dry the pressure would be less and the apparatus would appear as in B. The pressure is still less than the pressure of the room and less than in the initial condition.

9. a) the kinetic energy is the same in the three flasks because the temperature is the same. b) Flask Y because it has the highest pressure. c) Flask Z because H2 molecules have the lowest molar mass and the fastest velocity d) Flask Y because it has the largest mass and all the volumes are identical.

10. When the gases react, B is limiting and totally consumed. Only 2 moles of A react. Thus when the reaction is complete the gases present are 4 mol A, 0 mol B, 3 mol C in a total volume of 16 liters. The total pressure in the apparatus is 9.80 atm.

Section B: Multiple Choice

11. Which of the following represents the largest gas pressure?

    a) 5.0 torr   b) 5.0 mm Hg   c) 5.0 atm   d) 5.0 kPa   e) 5.0 bar

12. Which of the following represents the small gas pressure?

    a) $1.8 \, lb/in^2$ b) 1.8 torr   c) 1.8 atm   d) 1.8 kPa   e) 1.8 bar

13. Rank 355 torr, 0.524 atm and 0.513 bar in increasing order.

    a) 355 torr < 0.524 atm < 5.13 bar   b) 0.524 atm < 355 torr < 5.13 bar

    c) 5.13 bar < 355 torr < 0.524 atm   d) 355 torr < 5.13 bar < 0.524 atm

    e) 5.13 bar < 355 torr < 0.524 atm

14. Two identical manometers in the same laboratory at 760 torr are filled mercury. One contains Gas A and the other Gas B.   What is the pressure of the gases?

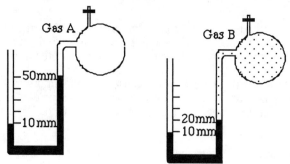

    a) Gas A = 750 torr, Gas B=720 torr   b) Gas A = 800 torr, Gas B=770 torr

    c) Gas A = 770 torr, Gas B=750 torr   d) Gas A = 720 torr, Gas B=750 torr

    e) Gas A = 710 torr, Gas B=740 torr

15. What are standard temperature and pressure conditions for gases?

    a) $0^{\circ}C$ and 0 torr   b) 0 K and 760 torr   c) $-273 \, ^{\circ}C$ and 1 atm

    d) $0^{\circ}C$ and 760 torr   e) $0^{\circ}C$ and 1 torr

16. If the volume of a confined gas is doubled while the temperature remains constant, what change (if any) would be observed in the pressure?
    a) It would be half as large.   b) It would double.
    c) It would be four times as large.   d) It would be one/fourth as large.
    e) It would remain the same.

17. A given mass of gas in a rigid container is heated from $100°C$ to $500°C$. Which of the following responses best describes what will happen to the pressure of the gas?
    a) The pressure will decrease by a factor of five.
    b) The pressure will increase by a factor of five.
    c) The pressure will increase by a factor of about two.
    d) The pressure will increase by a factor of about eight.
    e) The pressure will increase by a factor of about twenty-five.

18. Which of the following has the most molecules?
    a) 1.00 L of CH_4 at $0°C$ and 1.00 atm
    b) 1.00 L of N_2 at $0°C$ and 1.00 atm
    c) 1.00 L of O_2 at $20°C$ and 1.00 atm
    d) 1.00 L of CO_2 at $50°C$ and 1.25 atm
    e) 1.00 L of CO at $0°C$ and 1.25 atm

19. What volume of CH_4 at $0°C$ and 1.00 atm contains the same number of molecules as 0.50 L of N_2 measured at $27°C$ and 1.50 atm?
    a) 0.37 L   b) 0.46 L   c) 0.68 L   d) 0.50 L   e) 0.82 L

20. What volume of SO_2 at $25°C$ and 1.50 atm contains the same number of molecules as 2.00 L of Cl_2 measured at $0°C$ and 1.00 atm?
    a) 0.68 L   b) 1.22 L   c) 1.45 L   d) 1.83 L   e) 2.18 L

21. If 5.42 L of air measured at 735 torr and 23°C is heated to 35°C in the same container, what is the new pressure?
    a) 13.4 torr   b) 333 torr   c) 414 torr   d) 520 torr   e) 765 torr

22. If 3.0 L of helium at 20.0°C is allowed to expand to 4.4 L, with the pressure remaining the same, what is the new temperature?
    a) 157°C   b) 430°C   c) 702°C   d) –30.°C   e) – 55°C

23. A 4.5 L flask of Ar at 23°C and 734 torr is heated to 55°C. What is the new pressure?

    a) 366 torr   b) 935 torr   c) 1.25 torr   d) 1.07 atm   e) 2.58 atm

24. An air compressor reduced a sample of helium originally at 25°C and 740 torr to 6.75 liters at 42.0 atm and 85°C. What was the original volume of the helium?

    a) 85.6 liters   b) 35.9 liters   c) 242 liters   d) 319 liters   e) 350 liters

25. At what temperature will 5.0 grams CO_2 exerts a pressure of 815 torr in a 20.0 Liter cylinder?

    a) 134 K   b) 176 K   c) 238 K   d) 337 K   e) 400 K

26. What volume will a mixture of 0.200 mole N_2 and 0.500 mole He occupy at 0.944 atm and 15.0°C?

    a) 0.913 liters   b) 5.00 liters   c) 12.5 liters   d) 15.7 liters   e) 17.5 liters

27. When 7.0 grams of helium and 14.0 grams of argon were mixed in a flask the pressure was measured as 712 torr. What is the partial pressure of the helium?

    a) 593 torr   b) 356 torr   c) 833 torr   d) 1070 torr   e) 1420 torr

28. What pressure (in atmospheres) is exerted by 82.5 grams of CH_4 in a 75.0 liter container at 35.0°C?

    a) 0.197 atm   b) 0.339 atm   c) 1.73 atm   d) 2.57 atm   e) 27.8 atm

29. At STP, what is the volume of a mixture of gases containing 1.0 mol Ar, 1.5 mol He, and 2.5 mol Kr ?

    a) 5.0 liters   b) 17.5 liters   c) 22.4 liters   d) 112 liters   e) 345 liters

30. A mixture of the gases neon and krypton is in a 2.00 liter container. The partial pressure of the neon is 0.40 atm and the partial pressure of the krypton is 1.20 atm. What is the mol fraction of neon?

    a) 0.20   b) 0.25   c) 0.33   d) 0.60   e) 0.80

31. When 0.34 mols of He are mixed with 0.51 mols of Ar in a flask, the total pressure in the flask is found to be 5.0 atm. What is the partial pressure of Ar in this flask?

    a) 0.85 atm   b) 1.5 atm   c) 2.0 atm   d) 3.0 atm   e) 5.0 atm

32. When 20.0 mL of $SO_{2(g)}$ and 20.0 mL of $Cl_{2(g)}$ are allowed to react according to the equation below, what is the total volume of all gases after the reaction provided all gases are at the same temperature and pressure?

$$SO_{2(g)} + 2\,Cl_{2(g)} \longrightarrow OSCl_{2(g)} + Cl_2O_{(g)}$$

    a) 20.0 mL   b) 26.6 mL   c) 30.0 mL   d) 40.0 mL   e) 66.6 mL

33. Nitrogen and hydrogen gases react to form ammonia gas:

$$N_2\,(g) + 3\,H_2\,(g) \longrightarrow 2\,NH_3\,(g)$$

    At a certain temperature and pressure, 7.0 L of N_2 is reacted with 21.0 L of H_2. If all the N_2 and H_2 are consumed, what volume of NH_3 at the same temperature and pressure will be produced?

    a) 7.0 L   b) 14 L   c) 21 L   d) 28 L   e) 9.3 L

34. Which of the following gases has the greatest density at $0^\circ C$ and 1 atm?

    a) N_2 b) O_2 c) F_2 d) Ne   e) CO

35. Calculate the density of SO_3 gas at $35^\circ C$ and 715 torr.

    a) 0.0285 g/L   b) 1.43 g/L   c) 2.15 g/L   d) 2.98 g/L   e) 3.57 g/L

36. What is the density of CH_4 at $200^\circ C$ and 0.115 atm?

    a) 0.0475 g/L   b) 0.0716 g/L   c) 0.542 g/L   d) 0.870 g/L   e) 2.09 g/L

37. What is the density of SO_2 at $83^\circ C$ and 1.3 atm?

    a) 3.73 g/L   b) 2.86 g/L   c) 1.72 g/L   d) 1.10 g/L   e) 0.582 g/L

38. Air has a density of 1.29 g/L at $0^\circ C$ and 1.00 atm. What is the density of air at 655 torr and $-20.0\,^\circ C$?

    a) 1.69 g/L   b) 2.89 g/L   c) 4.52 g/L   d) 17.1 g/L   e) 22.1 g/L

39. At what temperature will the density of Ar be 2.66 g/L when the pressure is 1.32 atm?
    a) $-230^{\circ}C$ b) $-74^{\circ}C$ c) $-32^{\circ}C$ d) $17^{\circ}C$ e) $43^{\circ}C$

40. What is the molar mass of a gas which has a density of 1.30 g/L measured at $27.07^{\circ}C$
    and 0.400 atm?
    a) 38.0g/mol   b) 48.0 g/mol   c) 61.5 g/mol
    d) 80.1 g/mol   e) 97.5 g/mol

41. What is the molar mass of a gas which has a density of 1.83 g/L measured at $27.0^{\circ}C$ and
    0.538 atm?
    a) 25.0 g/mol   b) 38.0g/mol   c) 45.8 g/mol
    d) 75.4 g/mol   e) 83.8 g/mol

42. What is the chemical formula of a gas if it exerts a pressure of 1.40 atm at $27.0^{\circ}C$ and has a
    density of 1.82 g/L also at $27.0^{\circ}C$.
    a) CO_2 b) CO c) CH_4 d) O_2 e) N_2

43. What is the chemical formula of a gas if it exerts a pressure of 680 torr at $27.0^{\circ}C$ and has a
    density of 1.60 g/L also at $27.0^{\circ}C$.
    a) C_2H_6 b) CO_2 c) NO d) F_2 e) CF_4

44. The density of a gas composed only of carbon and chlorine is 6.24 g/L at $22.5^{\circ}C$ and 748
    torr. If the gas is 7.81% carbon, find the molecular formula for this gas.
    a) C_4Cl b) CCl_4 c) C_2Cl_2 d) C_2Cl_4 e) C_2Cl_6

45. The density of a gas composed only of carbon and fluorine is 5.61 g/L at $22^{\circ}C$ and 748 torr.
    If the gas is 17.40% carbon, find the molecular formula for this gas.
    molar mass is 138.02
    a) CF_2 b) CF_4 c) C_2F_3 d) C_2F_4 e) C_2F_6

46. What volume of O_2, collected at $22.0^{\circ}C$ and 728 mm Hg would be produced by the
    decomposition of 8.15 g $KClO_3$?
    $$2\ KClO_{3\ (s)}\ \ ----->\ \ 2\ KCl_{\ (s)}\ \ +\ 3\ O_{2\ (g)}$$
    a) 1.68 L   b) 1.12 L   c) 1.48 L   d) 2.23 L   e) 2.53 L

47. Gaseous iodine pentafluoride, IF_5, can be prepared by the reaction of solid iodine and gaseous fluorine according to the equation

$$I_{2(s)} \quad + \; 5\,F_{2(g)} \longrightarrow \quad 2\; IF_{5(g)}$$

What volume of F_2, measured at $37^{\circ}C$ and 705 torr, is needed to react completely with 350.0 grams of I_2?

 a) 378 L   b) 189 L   c) 75.6 L   d) 45.3 L   e) 37.9 L

48. What volume of O_2, measured at $25^{\circ}C$ and 733 torr, is needed for the complete combustion of 42.0 grams ethylene, C_2H_4?

 a) 38.0 L   b) 44.9 L   c) 62.7 L   d) 114 L   e) 341 L

49. What volume of O_2, measured at $27.0^{\circ}C$ and 0.750 atm, is needed for the complete combustion of 24.0 grams propylene, C_3H_8?

 a) 112 L   b) 89.3 L   c) 60.9 L   d) 53.6 L   e) 17.8 L

50. How many grams of CH_4 will react completely with 7.5 liters of O_2 measured at $150^{\circ}C$ and 0.850 atm in a combustion reaction?

 a) 1.47 g   b) 2.94 g   c) 1.84 g   d) 5.89 g   e) 0.011 g

51. A 9.56 gram sample of an ore contained iron pyrite, FeS_2, and non-reactive impurities. If 3.56 liters of SO_2 measured at $35^{\circ}C$ and 722 torr were produced, what was the % FeS_2 in the sample?

$$4\,FeS_{2(s)} + 11\,O_{2(g)} \longrightarrow \quad 2\,Fe_2O_{3(s)} + 8\,SO_{2(g)}$$

 a) 8.00%   b) 46.3%   c) 74.8 %   d) 84.0 %   e) 92.0%

52. If 50.0 g of HgO is decomposed to produce 2.80 liters of O_2 measured at $25^{\circ}C$ and 0.922 atm, what is the % yield.

$$2\;HgO\;(s) \longrightarrow \quad 2\;Hg\;(l)\; + \;O_2(g)$$

 a) 94.4%   b) 91.4 %   c) 67.3 %   d) 61.2%   e) 56.0%

53. If the following reaction produces 18.5 liters of NO at $22.0^{\circ}C$ and 735 torr, how many grams of I_2 are also produced?

$$6\;KI\;(aq)\; + \;8\;HNO_3(aq) \rightarrow \;6\;KNO_3\;(aq)\; + \;2\;NO(g)\; + 3\;I_2(s) + 4\;H_2O\;(l)$$

 a) 93.7 g   b) 125 g   c) 141 g   d) 187 g   e) 281 g

54. How many grams of gold are required to react completely with 17.6 liters of O_2 at 27.0°C and 715 torr in the presence of excess KCN and H_2O?

    4 Au (s) + 8 KCN(aq) + O_2(g) + 2 H_2O (l) ⟶ 4 K Au(CN)$_2$(aq) + 4 KOH(aq)

    a) 73.3 g   b) 132 g   c) 293 g   d) 530. g   e) 672 g

55. Calculate the volume of $Cl_{2(g)}$ produced at 815 torr and 15.0°C if 6.75 grams of $KMnO_4$ are added to 255 mL of 0.115 M HCl. The equation for the reaction is

    2 $KMnO_{4(s)}$ + 16 HCl(aq) ⟶ 2 $MnCl_{2}$(aq) + 2 KCl(aq) + 5 $Cl_{2(g)}$ + 8 $H_2O_{(l)}$

    a) 0.202 L   b) 0.470 L   c) 0.647 L   d) 0.942 L   e) 2.35 L

56. If 8.0 mol of the gas A is introduced in a 60.0 liter tank with 1.0 mol B at 0°C and 1.0 atm., they react according to the equation

    2 A (g) + B(g) ⟶ 3 C(g) + D(g)

    The temperature is held constant. What is the pressure in the flask when the reaction is complete?

    a) 1.12 atm   b) 1.49 atm   c) 2.99 atm

    d) 3.36 atm   e) 3.74 atm

57. Which statement about kinetic energy is true?

    a) All particles moving with the same velocity have the same kinetic energy.

    b) All particles at the same temperature have the same kinetic energy.

    c) All particles having the same kinetic energy have the same mass.

    d) As the kinetic energy of a particle is halved its velocity is also halved.

    e) As the velocity of a particles is doubled, the kinetic energy increases by a factor of four.

58. Equal masses of SO_2 and N_2 at the same temperature and pressure

    a) contain the same number of molecules

    b) have the same average kinetic energy

    c) have the same volume

    d) have the same density

    e) all of these

59. A plot of the relative number of gas molecules versus molecular velocity at 1000 K is given. Which of the accompanying plots would represent the same gas at 300 K?

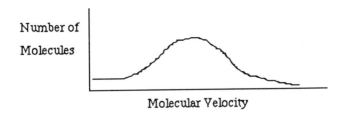

Number of Molecules

Molecular Velocity

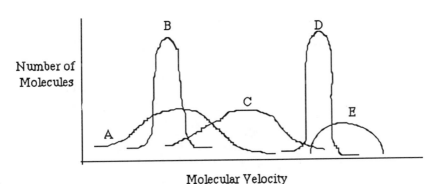

Number of Molecules

Molecular Velocity

60. Which of the following gases effuses at the highest rate?
    a) N_2 b) O_2 c) F_2 d) Ne   e) CO

61. At a particular temperature, which of the following molecules has an average velocity closest to that of ethylene, C_2H_4, at the same temperature?
    a) N_2 b) CO_2 c) NO_2 d) O_2 e) CH_4

62. Non-ideal behavior for a gas is most likely to be observed under conditions of
    a) standard temperature and pressure
    b) low temperature and high pressure
    c) low temperature and low pressure
    d) high temperature and high pressure
    e) high temperature and low pressure

Chapter 12 : Answers to Multiple Choice

11. e	21. e	31. c
12. b	22. a	32. d
13. d	23. d	33. b
14. d	24. c	34. c
15. d	25. b	35. d
16. a	26. e	36. a
17. c	27. a	37. a
18. e	28. c	38. a
19. c	29. d	39. c
20. c	30. b	40. d

41. e	51. d	61. a
42. d	52. b	62. b
43. b	53. e	
44. b	54. d	
45. e	55. a	
46. e	56. e	
47. b	57. b	
48. d	58. b	
49. b	59. b	
50. a	60. d	

Chapter 13
Bonding and Molecular Structure:
Intermolecular Forces, Liquids, and Solids

Section A: Free Response

1. Explain why the molar enthalpy of vaporization for H_2O is greater than the molar enthalpy of vaporization for Ne.

2. Of the molecules listed with their molar masses, only H_2O is a liquid at room conditions while the others are gases. Explain.

Compound	Molar Mass
CH_4	16
H_2O	18
CO	28
NO_2	46

3. A 0.200 gram sample of liquid water is placed in a closed 0.500 liter container at room temperature (approx. 20°C) and warmed to 70.0°C. Does any liquid remain when the temperature is held at 70.0°C (343 K)? Explain your choice based on your calculations.

Vapor Pressure of Water	
Temp(°C)	Pressure(atm)
0.0	0.006
40.0	0.073
70.0	0.308
90.0	0.692
100.0	1.000

4. Discuss the principal intermolecular force that is observed as each of the molecules H_2O, $(CH_3)_3N$, and N_2 in the gaseous state is cooled.

5. Rank the gases, O_2, NH_3 and Ne in and increasing order of solubility in water. Discuss the intermolecular forces which affect the solubility.

6. If the viscosities of the pure liquids A, B, and C are measured and it is found that the viscosity of A is greater than the viscosity of B which in turn is greater than the viscosity of C, then it is most likely that the substance with the highest surface tension is

7. Two compounds, A and B, have the same chemical composition. The vapor pressure of A is 500 torr at 25°C and the vapor pressure of B is 500 torr at –5°C. Which compound has the higher normal boiling point ? Sketch a vapor pressure graph to explain your choice.

8. Construct a phase diagram for Compound Z which has the following properties. a) the solid is more dense than the liquid b) the normal melting point is 40oC   c) the normal boiling point 84°C.   On your phase diagram insert the points A,B,C, and D such that the line AB represents sublimation, BC represents condensation, CD represents freezing.

9. Calculate the density of lithium metal at 20.0°C if it forms a body centered cubic unit cell and the radius of the lithium atom is 1.52 x 10⁻⁸ cm.

10. Use the phase diagram for Compound X to explain
    a) what would be observed as a sample of Compound X at -80.°C and 2.0 atm is heated to –55.°C and 2.0 atm.
    b) what would next be observed if the pressure is increased from 2.0 atm to 6.0 atm and the temperature maintained at –55°C.

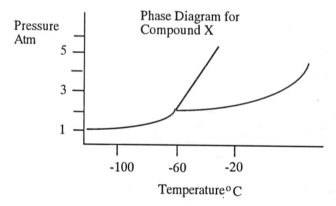

Key Concepts for Free Response

1. The molar masses are nearly the same so the important factor is the intermolecular forces between molecules of water which makes it difficult for them to escape from the liquid state. The hydrogen of one molecule has an attraction for the oxygen of another molecule which must be overcome before vaporization can occur.

2. The strong hydrogen bonding between water molecules makes them aggregate rather than stay independent as in the gaseous state at room temperature. All the other molecules do not have the ability to form hydrogen bonds.

3. Yes all of the water has vaporized. The pressure calculated from PV=nRT is 0.619 atm which is greater than the 0.308 atm that saturates the space above a water sample at $70^{\circ}C$.

4. As gaseous H_2O is cooled hydrogen bonding between the molecules occurs and they aggregate together to form a liquid. The bond is from the hydrogen of one molecule and the oxygen of another molecule of H_2O. For $(CH_3)_3N$ no hydrogen bonding can occur so the intermolecular force is the dipole-dipole interaction between the unshared pair of electrons on nitrogen and the more positive end of another molecule of $(CH_3)_3N$. In N_2 the only force is the weak induced dipole-induced dipole which accounts for the very low boiling point of liquid N_2.

5. $Ne < O_2 <<< NH_3$ Neon is least soluble because it has only an induced dipole-dipole interaction with water. O_2 is slightly more soluble because the diatomic molecule allows for some charge separation. NH_3 is very soluble in water because the unshared pair of electrons on the nitrogen forms hydrogen bonds with the hydrogen of water.

6. Viscosity is a resistance to flow. A is very viscous and does not flow easily. Thus it has strong intermolecular forces. Similarly it would have a high surface tension.

7. Compound A has the higher boiling point because at $25^{\circ}C$ it is still a liquid and Compound B would probably be a gas at that point.

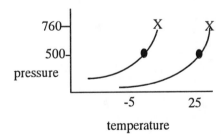

8. Phase Diagram

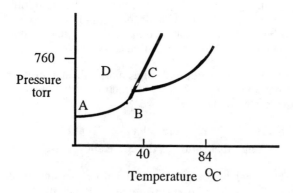

9. The density of lithium is 0.533 g/cm^3

10. a) Cmpd X would sublime from a solid at -80 °C to a gas at -55°C and constant pressure. No liquid would be observed. b) Cmpd X as a gas would liquefy and then solidify as the pressure is increased at -55oC. The solid compound is more dense than the liquid and would be observed on the bottom of the container as the process occurs.

Section B: Multiple Choice

11. When water molecules in the liquid state enter the gaseous state we can say that
    a) the *inter*molecular forces have weakened.
    b) the *intra*molecular forces have weakened.
    c) the intermolecular forces have strengthened
    d) the intramolecular forces have strengthened.
    e) both the inter and intramolecular forces have weakened.

12. Which one of the following substances would exhibit dipole-dipole intermolecular forces?
    a) CCl_4 b) Cl_2 c) N_2 d) NCl_3 e) CH_4

13. The forces that exist between noble gas atoms in the liquid and solid state are
    a) dipole/dipole   b) ion/dipole
    c) induced dipole/induced dipole   d) hydrogen bonds
    e) ion/ion

14. At room temperature, which of the following compounds has the strongest interparticle forces?
    a) CO_2 b) H_2O c) NaCl   d) CH_3CH_3 e) CH_3Cl

15. Which of the following would probably have the highest boiling point?
    a) Ar   b) He   c) Kr   d) Ne   e) Xe

16. Of the gases, Ne, N_2, O_2, Kr, and CH_4, which one would you expect to be easiest to liquefy?
    a) Ne   b) N_2 c) O_2 d) Kr   e) CH_4

17. Which of the following would probably have the lowest boiling point?
    a) CH_4 b) SiH_4 c) PH_3 d) AsH_3 e) NH_3

18. Which of the following molecules would exhibit hydrogen bonding in the liquid state?
    a) CH_4 b) H_2 c) NH_3 d) O_2 e) H_2S

19. Hydrogen bonding is a significant intermolecular force in the liquid state of which of the following substances?

a) HOCl   b) NF_3 c) H_2 d) CH_4 e) SiH_4

20. Which of the following would probably have the highest melting point?

a) Cl_2 b) NaCl   c) $MgCl_2$ d) CCl_4 e) CH_4

21. Which of the following would probably have the highest melting point?

a) LiBr   b) LiF   c) LiCl   d) NaCl   e) NaI

22. Which of the following would probably have the lowest melting point?

a) LiBr   b) $CaCl_2$ c) NaBr   d) KBr   e) MgF_2

23. Which of the following substances would probably have the highest melting point?

a) CaO   b) Br_2 c) CO_2 d) HCl   e) H_2O

24. Which of the following substances has the highest melting point?

a) B_2H_6 b) NaCl   c) CBr_4 d) $BaCl_2$ e) HBr

25. Which of the following molecules is most polar?

a) Cl_2 b) HCl   c) NO   d) IBr   e) H_2

26. Rank the compounds NH_3, CH_4, and CO in order of increasing boiling point:

a) $NH_3 < CH_4 < CO$ b) $CH_4 < NH_3 < CO$ c) $NH_3 < CO < CH_4$
d) $CH_4 < CO < NH_3$ e) $CO < NH_3 < CH_4$

27. If KBr, C_2H_5OH, C_2H_6 and He are arranged in order of increasing boiling point the list is

a) $He < C_2H_5OH < C_2H_6 < KBr$ b) $KBr < C_2H_6 < C_2H_5OH < He$
c) $He < C_2H_6 < C_2H_5OH < KBr$ d) $KBr < C_2H_6 < He < C_2H_5OH$
e) $C_2H_5OH < C_2H_6 < He < KBr$

28. Liquids with a high surface tension tend to have

    a) high intermolecular forces   b) high adhesive forces

    c) high intramolecular forces   d) low cohesive forces

    e) no cohesive forces

29. As ice melts what bonds, if any, are broken?

    a) one of the hydrogen and oxygen bonds within a water molecule

    b) both of the hydrogen-oxygen bonds within a water molecule

    c) hydrogen bonds between water molecules

    d) all of the above

    e) no bonds are broken

30. The boiling points of NH_3, PH_3, AsH_3, and SbH_3 follow a periodic trend except for

    a) NH_3 which has an unexpectedly high boiling point.

    b) NH_3 which has an unexpectedly low boiling point.

    c) SbH_3 which has an unexpectedly high boiling point.

    d) SbH_3 which has an unexpectedly low boiling point.

    e) AsH_3 which has an unexpectedly high boiling point.

31. The trends in boiling points of HF, HCl, HBr and HI respectively are depicted as X on the graph. What conclusion can be drawn concerning intermolecular forces?

Boiling
Point

             Molar Mass

    a) HF has very weak induced dipole-induced dipole interactions.

    b) HF has strong hydrogen bonding.

    c) HCl has strong induced dipole-induced dipole interactions.

    d) HI has a strong dipole/dipole interactions.

    e) HI has strong hydrogen bonding.

32. The trends in boiling points of CH_3CH_3, CH_3CH_2Cl, CH_3CH_2Br, and CH_3CH_2I respectively are shown on the graph. Why is the trend not linear?

Boiling
Point

Molar Mass

a) CH_3CH_3 does not have a permanent dipole.
b) CH_3CH_3 has strong hydrogen bonding.
c) CH_3CH_2Cl has strong induced dipole-induced dipole interactions.
d) CH_3CH_2Cl does not have a permanent dipole.
e) CH_3CH_2I has the largest molar mass and the strongest hydrogen bonding.

33. The equation which represents the number of atoms in a face-centered cubic unit cell is
    a) $8(1/8) + 4(1/2)$ b) $4(1/4) + 4$ c) $6(1/4) + 6(1/2)$
    d) $8(1/8) + 4(1/4) + 2(1/2)$ e) $(8)(1/8) + 6(1/2)$

34. An atom that is shared equally between eight cubic unit cells is called
    a) an edge atom   b) a face atom   c) a corner atom
    d) a diagonal atom   e) a central atom

35. A metal forms a simple cubic unit cell with a volume, V. Which of the following expressions allows one to calculate the radius of the atom?
    a) $\sqrt[3]{V}$

    b) $2\sqrt[3]{V}$

    c) $\sqrt[3]{2}\,\sqrt[3]{\dfrac{V}{2}}$

    d) $\sqrt[3]{\dfrac{V}{4}}$

    e) $\sqrt[3]{\dfrac{V}{2}}$

36. The volume of a body-centered cubic unit cell (bcc) of tungsten is 3.31×10^{-23} cm^3. The density of tungsten is
    a) 36.8 g/cm^3 b) 18.4 g/cm^3 c) 9.22 g/cm^3
    d) 5.55 g/cm^3 e) 2.44 g/cm^3

37. Copper crystallizes in a face-centered cubic unit cell. If the atomic radius of copper is 1.27 x 10^{-8} cm, calculate the density of copper metal in g/cm^3.
    a) 19.7 g/cm^3 b) 5.40 g/cm^3 c) 6.04 g/cm^3
    d) 9.11 g/cm^3 e) 36.4 g/cm^3

38. The density of calcium is 1.61 g/cm^3 and the edge of the calcium unit cell is 5.49 x 10^{-8}cm. Calculate the number of atoms in the unit cell.
    a) 1   b) 2   c) 4   d) 6   e) 8

39. The element titanium has a density of 4.25 g/cm^3 and forms a body-centered cubic (bcc) unit cell. Calculate the length of the edge of the unit cell.
    a) 1.67 x 10^{-8} cm   b) 3.34 x 10^{-8}cm   c) 4.03 x 10^{-8} cm
    d) 6.68 x 10^{-8} cm   e) 6.61 x 10^{-9} cm

40. Nickel crystallizes in a face-centered cubic array of atoms. The edge of the unit cell is 3.51 x 10^{-8} cm. What is the density of this metal?
    a) 18.0 g/cm^3 b) 2.25g/cm^3 c) 9.01 g/cm^3
    d) 0.111 g/cm^3 e) 36.1 g/cm^3

41. The edge of a face-centered cubic unit cell (fcc) of neon is 4.52 x $10^{-8}cm^3$. The density of neon is
    a) 0.361 g/cm^3 b) 0.724 g/cm^3 c) 1.45 g/cm^3
    d) 2.90 g/cm^3 d) 3.70 g/cm^3

42. Chromium crystallizes in a body-centered cubic unit cell. In a bulk sample of chromium each atom is surrounded by how many "nearest neighbors"?
    a) 2   b) 4   c) 6   d) 8   e) 12

43. What fraction of an atom occupying a face position of a cubic lattice is part of the unit cell?
    a) 1   b) 1/2   c) 1/4   d) 1/6   e) 1/8

44. The figure at the right represents a unit cell where and • is copper on the faces and o is gold at the corners. The empirical formula of the alloy is

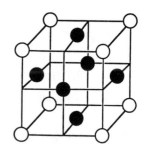

   a) $CuAu_2$ b) Cu_3Au c) Cu_6Au_8

   d) $CuAu_6$ e) Cu_6Au

45. In the unit cell at the right • represents element X within the cell and o represents element Y on the corners. The formula of the compound is

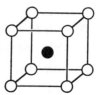

   a) XY   b) XY_2 c) $XY4_8$

   c) XY_6 d) XY_8

46. In the unit cell at the right • represents element A within the cell and o represents element B on the corners. The formula of the compound is

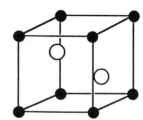

   a) A_2B b) AB_4 c) AB_6

   d) AB_2 e) AB

47. A metal fluoride crystallizes such that the fluoride ions occupy cubic lattice positions at the corners and on the faces while the 4 metal atoms occupy positions within the body of the unit cells. The formula of the metal fluoride is
   a) MF   b) MF_2 c) MF_3 d) M_4F_{14} e) MF_8

48. MgO has the same crystal structure as NaCl. How many oxide ions surround each Mg^{2+} ion in the solid state.
   a) 1   b) 2   c) 4   d) 6   e) 8

49. Which of the following properties of water can be attributed to hydrogen bonding?

a) high surface tension   b) high boiling point

c) high heat of vaporization   d) high specific heat

e) all of these

50. For a pure substance, the ΔH_{fusion} is known to be 15.0 kJ/mol. Which of the following is most probably the $\Delta H_{vaporization}$ for this substance?

a) –15.0 kJ/mol   b) + 15.0 kJ/mol   c) 0.0 kJ/mol

d) –45.0 kJ/mol   e) 45.0 kJ/mol

51. When 25 kJ of energy is absorbed by 50.0 g ice at $0°C$, what is observed?
Heat of fusion $_{H_2O}$ = 333 J /g;
Specific Heat $_{ice}$ =2.06 J /g·K; Specific Heat $_{water}$ = 4.18 J /g·K

a) the ice melts and the water temperature rises to $16.°C$.

b) the ice melts and the water temperature rises to $40°C$.

c) the ice melts and the water temperature rises to $65.°C$.

d) the ice melts and the water temperature rises to $80°C$.

e) the ice partially melts and the temperature of the mixture is $0°C$.

52. If 35 kJ of energy is removed from 75.0 grams H_2O at $20.0°C$ what is observed?
Heat of fusion $_{H_2O}$ = 333 J /g;
Specific Heat $_{ice}$ =2.06 J /g·K; Specific Heat $_{water}$ = 4.18 J /g·K

a) the H_2O freezes to solid ice at $0°C$.

b) the H_2O partially freezes to form a water/ice mixture at $0°C$.

c) the H_2O freezes to solid ice at –8.5$°C$.

d) the H_2O freezes to solid ice at –24$°C$.

e) the H_2O freezes to solid ice at – 47$°C$.

53. Methanol, CH_3OH, has a heat of vaporization of 39.2 kJ/mol and a density of 0.7914 g/mL. How much energy is needed to vaporize 350. mL of ethanol?

a) 1.08 x 10^4 kJ   b) 8.86 x 10^2 kJ   c) 428 kJ

d) 652 kJ   e) 339 kJ

54. Which of the following processes (if any) is exothermic?
    a) solid ----> gas
    b) solid ----> liquid
    c) liquid ---> solid
    d) liquid ---> gas
    e) none of these

55. For a pure substance, the ΔH_{fusion} is known to be 15.0 kJ/mol. Which of the following is most probably the $\Delta H_{vaporization}$ for this substance?
    a) –15.0 kJ/mol
    b) –45.0 kJ/mol
    c) 0.0 kJ/mol
    d) +15.0 kJ/mol
    e) +45.0 kJ/mol

56. A chemist sets up two beakers of distilled water under the same room conditions in a laboratory. One beaker is boiling vigorously, the other beaker is boiling gently. Which of the following statements is true?
    a) The temperature of the vigorously boiling water is higher.
    b) The temperature of the gently boiling water is higher.
    c) The temperature of the water in both beakers is the same
    d) The boiling points of the water in the two beakers must be different.
    e) The temperature in the vigorously boiling water is not uniform.

57. As the pressure on a pure substance is increased at constant temperature, a phase transition (if any) which is *not* observed would be
    a) gas -----> solid
    b) liquid -----> solid
    c) solid -----> gas
    d) gas -----> liquid
    e) all of these could be observed

58. The greatest change in energy for a substance is seen with which process?
    a) vaporization
    b) condensation
    c) fusion
    d) sublimation
    e) melting

59. Which of the following exhibit NO CHANGE IN TEMPERATURE when 27 Joules of heat energy is removed?
    a) 100 g $H_2O(s)$ at –10°C
    b) 100 g $H_2O(l)$ at 0°C
    c) 100 g $H_2O(l)$ at 75°C
    d) 100 g $H_2O(g)$ at 110°C
    e) 100 g $H_2O(s)$ at 0°C

60. Using the phase diagram for Substance Z, what changes would be observed it the conditions are changes from A to B to C to D in that order?

a) melting, vaporization, deposition

b) vaporization, freezing, sublimation

c) sublimation, freezing, melting

d) freezing, sublimation, vaporization

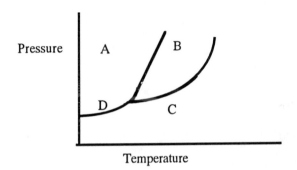

61. A substance at a temperature greater than its critical temperature is called

    a) a solid   b) a vapor   c) a rheostatic liquid

    d) a supercritical fluid   e) an hydraulic fluid

62. Which of the following is an example of a network solid?

    a) SiO_2 b) MgO c) P_4 d) $NaCl$ e) I_2

63. A method useful for preparing synthetic diamonds by decomposing methane and hydrogen at about $2000\,^\circ C$ is called

    a) chemical vapor deposition   b) electrolysis   c) sublimation

    d) scanning tunneling microscopy   e) infrared spectroscopy

Chapter 13 : Answers to Multiple Choice

11. b	21. b	31. b
12. d	22. d	32. a
13. c	23. a	33. e
14. c	24. d	34. c
15. e	25. b	35. e
16. d	26. d	36. b
17. a	27. c	37. d
18. c	28. a	38. c
19. a	29. c	39. b
20. c	30. a	40. c

41. c	51. d	61. d
42. d	52. d	62. a
43. b	53. e	63. a
44. a	54. c	
45. a	55. e	
46. d	56. c	
47. a	57. c	
48. d	58. d	
49. e	59. b	
50. e	60. a	

Chapter 14
Solutions and Their Behavior

Section A: Free Response

1. When solid pellets of sodium hydroxide are dissolved in water at room temperature, the resulting solution is much warmer than room temperature. Explain what processes are involved in forming this solution and why the solution is warmer.

2. When a warm glass rod is inserted in a cold freshly opened carbonated beverage, bubbles appear on the rod. What are the bubbles? Explain why they are formed on the rod.

3. Which solution of KBr, 5.0 ppm or 0.05%, is more dilute? Explain.

4. Of the concentration units, mole fraction, parts per million (ppm) or molarity, which one is temperature dependent?   Explain.

5. A research team prepared a new compound which has a molar mass of 175. A novice chemist then prepared a solution of a new compound by dissolving 0.80 g in a small amount of water and diluting to a volume of 50.0 mL. A supervisor asked for the **molality** of the solution. Does the chemist have enough information to determine the molality? If so, calculate the molality.   If not, explain what additional information could be determined by the chemist to complete the task in a short time.

6. Formaldehyde, CH_2O, is sold commercially as a 37.0% by weight aqueous solution which has a density of 1.08 g/cm^3. What is the vapor pressure of the solution at 20.0°C, if the vapor pressure of H_2O is 17.5 torr at 20.0°C?

7. The phase diagram for pure Solvent A is given. Draw a dotted line (---) appropriate for the vapor pressure of a solution of a non-volatile solute dissolved in Solvent A. Using your diagram estimate the normal boiling point and freezing point for your solution. Explain the trend observed.

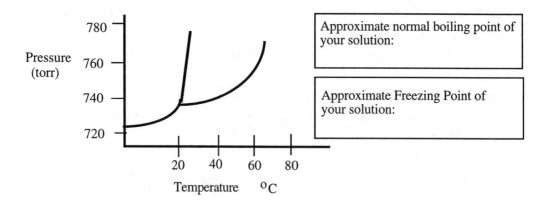

8. Consider the schematic figure below in which the top layer is representative of the non-volatile solute, ♦, dissolved in the solvent, o. Given that the vapor pressure of the pure solvent is 25.4 torr at room temperature, what is the expression for the vapor pressure of the solution at room temperature. Explain.

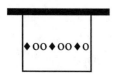

9. Compare the freezing points of two solutions, 0.3 m NH_4NO_3 and 0.3 m Na_2SO_4. Both solutes are non-volatile. Explain.

10. Equal volumes of two aqueous solutions, 5 M NaCl and 5 M MgCl₂, are placed in compartments separated by a semipermeable membrane in the apparatus the figure. Sketch the appearance of the system after equilibrium has been achieved. Explain fully.

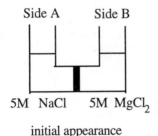

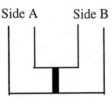

11. A compound is 64.87% carbon, 2.73% hydrogen and 32.41% oxygen. It is non-dissociating in solution.   When 3.14 g is dissolved in 35.0 g of camphor, the solution freezes at 167.5°C? Calculate the molecular formula of the compound. (Constants for pure camphor: freezing point 179.5°C; K_{fp} -39.7°C/m)

12. A solution is prepared by dissolving 3.500 g of the non-volatile Compound Z in 40.0 g of benzene. The boiling point of the solution is 84.34°C. The normal boiling point of benzene is 80.10°C and K_{bp} = +2.53°C/m . What is the molar mass of the Compound Z?

13. What laboratory measurements can be used to distinguish a true solution from a colloid? Explain using appropriate examples.

14. Explain how the structure of a surfactant changes the surface properties of water.

15. When milk ferments, the hydrogen ion content of the solution increases. Explain how this causes the proteins which were in colloidal suspension to solidify as curds.

Chapter 14: Answers to Free Response

1. To form a solution energy must be added to break the crystal of the Na^+ and OH^-. As the water molecules surround the ions, energy is given off. Since the overall process is exothermic, we conclude that more energy is given off as the ions are solvated than is required to break the crystal.
2. The bubbles are carbon dioxide gas on the warm rod. This illustrates that gases are less soluble at higher temperatures which would be expected because there is more energy in the system and it is easier for the gases to escape from the solution.
3. The 0.05% KBr solution is more dilute because it has 5.0×10^{-4} g/L while the 5.0 ppm has 5.0×10^{-3} g/L.
4. Molarity can change with temperature because the volume of the solution is measured. At higher temperatures the volume would be greater and the molarity would decrease slightly.
5. The molality cannot be calculated based on the molarity. The chemist need to determine the density of the solution by weighing a sample (e.g., 10.00 mL). Then the volume of solvent can be determined by subtracting the mass of the solute from the mass of the solution.
6. Mol fraction solvent = 0.741 and vapor pressure of the solution is 12.9 torr.
7. Normal boiling point must be > than 60°C and freezing point < 20°C.   Figure should show dotted lines higher than the boiling point equilibrium line and lower than the freezing point line in the same general shape as in the figure.

8. 15.9 torr. $P_{solution} = (5/8)25.4$ torr. A solution has a lower vapor pressure than the pure solvent because it requires more energy for the solvent particles to vaporize. The surface contains solute particles which interfere with the vaporization process. The mol fraction of the solvent (5/8) times the pressure of the pure solvent is the vapor pressure of the solution.
9. The vapor pressure of 0.3 m NH_4NO_3 is higher because only two particles (NH_4^+ and NO_3^-) are present on the surface of the solution. The 0.3 m Na_2SO_4 has 3 particles which reduce the vapor pressure of the solution more.
10. Water moves through the membrane from the NaCl side to the $MgCl_2$ side because fewer solute particles cause less pressure on the membrane. The solvent moves until the pressure on either side of the membrane is the same. At equilibrium, the NaCl side is lower and the $MgCl_2$ side is higher than in the initial condition.
11. Empirical formula $C_8H_4O_3$. From freezing point data, solution is 0.302 m and molar mass is 296. The molecular formula is $C_{16}H_8O_6$.
12. 52.2 g/mol
13. Light will be scattered by the larger size particles present in a colloid while light is not diffused in a true solution. Shining a laser or any intense light source will distinguish the two.   For example, a dilute starch solution may appear colorless to the eye, but light will be scattered in a starch solution. A sugar solution also appears colorless but light will not be scattered through it.
14. A surfactant (like soap) breaks up the hydrogen bonds between water molecules. This lessens the surface tension of the water and changes its properties. A surfactant has a polar end (charge such as the anion of an acid) which is attached to the water molecules so they no longer interact with each other as much.
15. The proteins in milk are a colloidal suspension of hydrophobic particles. When milk ferments, more ions are produced which breaks up the atmosphere surrounding the suspended particles and they can precipitate as curds.

Section B: Multiple Choice

16. What is the molality of a 7.80% by weight glucose ($C_6H_{12}O_6$) solution?

    a) 0.470 m b) 0.845 m c) 0.0432 m

    d) 0.0454m e) 0.0844 m

17. What is the molality of a 5.45% by weight Na_2SO_4 solution?

    a) 0.0383 m b) 0.0818 m c) 0.406 m

    d) 7.74 m e) 8.18 m

18. The mol fraction of NH_4Cl in a solution is 0.0311. What is its molality?

    a) 1.78 m b) 1.66 m c) 0.969 m

    d) 0.0983 m e) 0.562m

19. The mol fraction of sucrose ($C_{12}H_{22}O_{11}$) in a solution is 0.0584. What is its molality?

    a) 0.0621m b) 0.202 m c) 0.290 m

    d) 1.18 m e) 3.44 m

20. What is the weight % $CaCl_2$ of a 2.18 m $CaCl_2$ solution?

    a) 31.9%   b) 19.5%   c) 4.03%

    d) 2.41%   e) 0.287%

21. What is the weight percent $ZnCl_2$ of a 1.12 m $ZnCl_2$ solution?

    a) 1.52%   b) 7.55%   c) 13.2%

    d) 18.0%   e) 24.6%

22. What is the mol fraction KBr in a solution which is 14.3% by weight KBr?

    a) 0.0252   b) 0.0246   c) 0.167

    d) 0.888   e) 0.976

23. What is the mol fraction Na_2SO_4 in a solution which is 11.5% by weight Na_2SO_4?

    a) 0.0810   b) 0.0914   c) 0.0745

    d) 0.0173   e) 0.0162

24. If the mol fraction NaCl in a solution is 0.0175, what is the weight percent NaCl?
    a) 5.46%   b) 5.77%   c) 10.2%
    d) 11.5%   e) 17.7%

25. If the mol fraction NaCN in a solution is 0.0117, what is the weight percent NaCN?
    a) 3.12%   b) 3.22%   c) 5.73%
    d) 5.62%   e) 9.78%

26. What is the mol fraction $NaNO_3$ in a solution which is 2.15 m ?.
    a) 0.0180   b) 0.0268   c) 0.0373
    d) 0.09387   e) 0.0785

27. What is the mol fraction K_2SO_4 in a solution which is 3.24 m ?
    a) 0.0537   b) 0.0552   c) 0.0564
    d) 0.0584   e) 0.0598

28. What is the weight percent $FeCl_3$ in a solution with is 1.84 m ?
    a) 14.0%   b) 16.2%   c) 29.8%
    d) 25.2%   e) 23.0%

29. What is the weight percent NH_4Cl in a solution which is 3.15 m ?
    a) 1.68%   b) 5.35%   c) 14.4%
    d) 20.3%   e) 8.47%

30. A 1.34 M $NiCl_2$ solution has a density of 1.12 g/cm^3. What is the weight percent
    $NiCl_2$ of the solution?
    a) 1.73%   b) 8.64%   c) 15.5%
    d) 25.4%   e) 29.8%

31. A 1.34 M $NiCl_2$ solution has a density of 1.12 g/cm^3. What is the molality of the solution?
    a) 0.913 m b) 1.42 m c) 1.55 m
    d) 2.55 m e) 3.13 m

32. A 1.26 M $Cu(NO_3)_2$ solution has a density of 1.19 g/cm^3. What is the weight
    percent $Cu(NO_3)_2$ of the solution?
    a) 1.88%   b) 2.36%   c) 10.5%
    d) 14.3%   e) 19.9%

33. A 1.26 M $Cu(NO_3)_2$ solution has a density of 1.19 g/cm^3. What is the molality of
    the solution?
    a) 1.06 m b) 1.32 m c) 6.34 m
    d) 6.72 m e) 8.44 m

34. What is the mol fraction $Zn(NO_3)_2$ in a solution which is 1.45 M and has a density
    of 1.22 g/cm^3?
    a) 0.0269   b) 0.0276   c) 0.0951
    d) 0.291   e) 0.378

35. What is the weight percent $Zn(NO_3)_2$ in a solution which is 1.45 M and has a density
    of 1.22 g/cm^3?
    a) 6.44%   b) 7.65%   c) 24.5%
    d) 27.5%   e) 32.8%

36. What is the molarity of a 25.0% HCl solution if the density is 1.08 g/cm^3?
    a) 2.96 M   b) 2.70 M   c) 7.41 M
    d) 9.11 M   e) 5.49 M

37. Hydrobromic acid (HBr Molar Mass = 80.90) is commercially available in a 34.0
    mass percent solution which has a density of 1.31 g/cm^3. What is the molarity of
    the commercially available hydrobromic acid?
    a) 2.75 M   b) 4.45 M   c) 5.50 M
    d) 9.35 M   e) 10.2M

38. A solution of hydrogen peroxide is 30.0% H_2O_2 by weight and has a density of
    1.11 g/cm^3. The molarity of the solution is
    a) 9.79 M   b) 0.980 M   c) 7.94 M
    d) 8.82 M   e) 13.3 M

39. The maximum contamination level of arsenic ion in a water system is 0.050 parts per million. If the arsenic is present as $AsCl_3$, how many grams of arsenic chloride could be present in a system that contains 8.2×10^5 liters .

a) 0.55 g   b) 7.3 g   c) 41 g

d) 62 g   e) 98 g

40. A solution has a magnesium ion concentration of 1.2×10^3 ppm. If the magnesium ion is dissolved as $MgCl_2$, how many grams of $MgCl_2$ are in each liter of solution?

a) 1.14 g   b) 1.26 g   c) 4.70 g

d) 4.93 g   e) 5.84 g

41. A student prepared a solution containing 0.30 mol solute and 1.00 mole solvent. The mole fraction of solvent is

a) 1.30   b) 1.00   c) 0.77

d) 0.30   e) 0.23

42. Which measure of concentration is appropriate for the calculation of the vapor pressure of a solution?

a) mol fraction   b) molarity   c) molality

d) weight %   e) ppm

43. If 0.1 gram of a compound with a molar mass of 10,000 is dissolved in 50.0 grams of water, what colligative property is most appropriate to measure in order to determine the exact molar mass?

a) freezing point   b) vapor pressure   c) molarity

d) osmotic pressure   e) molality

44. A chemist knows the empirical formula of a new compound but not the molecular formula. What must be determined experimentally so that the molecular formula can be determined.

a) density   b) viscosity   c) % composition

d) melting point   e) molar mass

45. The amount of solvent (grams or moles) is known for each of the following solution concentrations EXCEPT

    a) molarity   b) molality   c) mass %

    d) χ e) Π

45. A volumetric flask is necessary for the preparation of which one of the following concentration measurements?

    a) molality   b) χ c) mass %

    d) molarity   e) Π

47. Which of the following substances is more soluble in hexane (C_6H_{14}) than in water?

    a) $CaCl_2$ b) CH_3OH c) NH_3

    d) $C_8H_{17}Cl$ e) CH_3Cl

48. If the pressure of a gas over a liquid increases, the amount of gas dissolved in the liquid will

    a) increase   b) decrease   c) remain the same

    d) have a higher vapor pressure   e) depend on the polarity of the gas

49. The liquids H_2O and CCl_4 are immiscible due to

    a) the strong intermolecular forces between H_2O molecules

    b) the strong intermolecular forces between CCl_4 molecules

    c) the strong dipole of the CCl_4 molecules

    d) the weak dipole of H_2O molecules

    e) the large difference in molar masses of H_2O and CCl_4

50. Which of the following would have the lowest vapor pressure?

    a) 1 m glucose ($C_6H_{12}O_6$)   b) 1 m $MgCl_2$ c) 1 m $NaNO_3$

    d) 1 m NaBr   e) pure H_2O

51. Which of the following would have the highest vapor pressure?

    a) pure H_2O b) 1 m glucose ($C_6H_{12}O_6$)   c) 1m $NaNO_3$

    d) 1m $MgCl_2$ e) 1 M $(NH_4)_2SO_4$

52. Which of the following would have the a boiling point closest to that of $1m$ NaCl?
    a) pure H_2O b) $1\ m$ sucrose $(C_{12}H_{22}O_{11})$ c)$1\ m$ $MgCl_2$
    d) $0.5\ m$ Cl_2 e) $1\ m$ NH_4NO_3

53. Which of the following solutions would have the a freezing point closest to that of a 1 molal solution of $CaCl_2$?
    a) $1\ m$ $CaSO_4$ b) $1m$ KBr   c) $1\ m$ Na_2SO_4
    d) $0.5\ m$ $SnCl_4$ e) pure H_2O

54. Which of the following would have the highest freezing point?
    a) $1\ m$ glucose $(C_6H_{12}O_6)$ b) $1\ m$ $MgCl_2$ c) $1\ m$ $NaNO_3$
    d) $1\ m$ $(NH_4)_2SO_4$ e) pure H_2O

55. Which of the following would have the lowest freezing point?
    a) pure H_2O b) $1\ m$ urea (CON_2H_4) c) $1\ m$ KCl
    d) $1\ m$ $NaNO_3$ e) $1\ m$ Na_2SO_4

56. When Compound Z is dissolved in water at room temperature, the solution is quite cold. Thus we may conclude that

    a) more energy is given off by the water molecules surrounding the ions than is absorbed by the breaking of the solid crystal lattice.
    b) less energy is given off by the water molecules surrounding the ions than is absorbed by the breaking of the solid crystal lattice.
    c) energy is absorbed by both the breaking of the solid crystal lattice and by the surrounding of the ions in solution.
    d) energy is released by both the breaking of the solid crystal lattice and by the surrounding of the ions in solution.
    e) the only energy change is for the breaking of the crystal lattice.

57. You need a solution that is $0.15m$ in ions. How many grams of $MgCl_2$ must you dissolve in 400. g of water? (Assume total dissociation of the ionic solid.)
    a) 0.060 g   b) 1.9 g   c) 5.7 g
    d) 17. g   e) 7.6 g

58. What is the freezing point of a solution containing 4.134 grams naphthalene (Molar
    Mass = 128.2) dissolved in 30.0 grams paradichlorobenzene? The freezing point of
    pure paradichlorobenzene is 53.0°C and the freezing point depression constant K_{fp}
    is 7.10°C/m.
    a) 52.0°C   b) 48.7°C   c) 45.4°C
    d) 17.6°C   e) 7.63°C

59. What is the molar mass of a non-dissociating solid if 5.48 grams dissolved in 35.0
    grams of benzene freezes at –1.39°C? The freezing point of pure benzene is
    5.50°C and the freezing point depression constant K_{fp} is –5.12°C/m.
    a) 118 g/mol   b) 189 g/mol   c) 245 g/mol
    d)412 g/mol   e) 609 g/mol

60. The freezing point of pure cyclohexane is 6.50°C. A solution is prepared by
    dissolving 0.500 g of a non-dissociating solute in 12.0 g of cyclohexane. The
    solution freezes at –2.44°C. The K_{fp} for cyclohexane is –20.0 °C/m. Calculate the
    molar mass of the solute.
    a) 93.2 g/mol   b) 112 g/mol   c) 128 g/mol
    d) 182 g/mol   e) 205 g/mol

61. What is the molar mass of a compound if 5.96 grams is dissolved in 25.0 grams of
    chloroform solvent to form a solution which boils at 66.50°C. The boiling point of
    pure chloroform is 61.70°C and the boiling point constant, K_{bp} is + 3.63°C/m.
    a) 112 g/mol   b) 132 g/mol   c) 180. g/mol
    d) 342 g/mol   e) 451 g/mol

62. What is the molar mass of a compound if 4.28 grams is dissolved in 25.0 grams of
    chloroform solvent to form a solution which has a boiling point of 64.00°C. The
    boiling point of pure chloroform is 61.70°C and the boiling point constant, K_{bp} is
    + 3.63°C/m..
    a) 35.4 g/mol   b) 67.5 g/mol   c) 168 g/mol
    d) 135 g/mol   e) 270. g/mol

63. The osmotic pressure of blood is 7.65 atm at 37°C. How many grams of glucose ($C_6H_{12}O_6$, Molar Mass = 148) are needed to prepare 1.00 liter of a solution for intravenous injection that has the same osmotic pressure as blood?

a) 3.00 g   b) 4.44 g   c) 25.4 g

c) 45.3 g   e) 54.1 g

64. When Solution A, 0.10M NaCl and Solution B, 0.20M NaCl are separated by a semipermeable membrane, what occurs during osmosis?

a) solvent molecules move from B into A

b) Na^+ and Cl^- ions move from B into A

c) the vapor pressure of A increases.

d) the molarity of A increases.

e) no change is observed.

65. When a liquid is dispersed in another liquid, the resulting mixture is called a(n)

a) gel   b) sol   c) emulsion   d) colloid   e) alloy

66. The hydrophobic portion of a soap molecule is

a) the long carbon chain   b) the carboxylic acid   c) the glycerol

d) the anion of the acid   e) the short carbon chain

Chapter 14: Multiple Choice Answers

16. a	26. c	36. c
17. c	27. b	37. c
18. a	28. e	38. a
19. e	29. c	39. e
20. b	30. c	40. c
21. c	31. b	41. c
22. b	32. e	42. a
23. e	33. b	43. d
24. a	34. a	44. e
25. a	35. c	45. a

46. d	56. b	66. a
47. d	57. b	
48. a	58. c	
49. a	59. a	
50. b	60. a	
51. a	61. c	
52. e	62. e	
53. c	63. e	
54. d	64. d	
55. e	65. c	

Chapter 15
Principles of Reactivity: Chemical Kinetics

Section A: Free Response

1. The kinetics of the reaction $2\,A + 3\,B \longrightarrow C + 2\,D$ are being studied. The data for changes in the concentrations of A and B are plotted in several ways below. Analyze the data, write the rate law and discuss the conclusion that can be made.

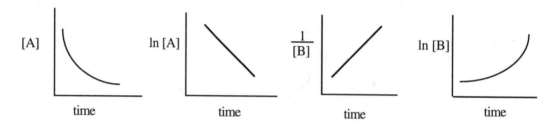

2. Experimentally what method would provide useful data to study the kinetics of the reaction
    $$CaCO_3(s) \longrightarrow CaO(s) + CO_2(g)$$

3. Explain why the initial reaction rate is determined for kinetic studies rather than the average reaction rate.

4. List those factors which influence the rate of a chemical reaction which are explicit or implicit in the Arrhenius equation, $k = Ae^{-Ea/RT}$.

5. The following mechanism has been proposed for the reaction of NO with H_2 which has the observed rate law, rate $= k\,[NO]^2$.

Step 1:	$NO(g) + NO(g) \longrightarrow N_2O_2(g)$
Step 2:	$N_2O_2(g) + H_2(g) \longrightarrow N_2O(g) + H_2O(g)$

    Discuss the plausibility of the mechanism and the relative rates of the two steps.

6. Experimentally how is the rate constant of a reaction determined?

7. Kinetic data for the following reaction was determined experimentally:

$$2 \ NO_{2(g)} \quad + \quad F_{2(g)} \quad \longrightarrow \quad 2 \ NO_2F_{(g)}$$

Experiment Number	Initial Conc. (mol/L) $[NO_2]_o$	Initial Conc. (mol/L) $[F_2]_o$	Initial rate of reaction (mol/L·s)
1	0.20	0.10	0.41
2	0.60	0.10	3.67
3	0.40	0.20	3.29
4	0.85	0.35	?

What is the rate law for the reaction? Calculate the initial rate of reaction for Experiment 4.

8. Contrast the method by which a catalyst changes the rate of reaction to the method by which temperature changes the rate of reaction.

9. Explain how a mechanism can be disproved by kinetic studies but not proven conclusively.

10. Discuss the major chemical reactions which the catalytic converter in an automobile engine promotes.

Key Concepts for Free Response

1. The reaction is first order with respect to A and second order with respect to B. The value of k, the rate constant can be calculated from the slope of each line. The numerical values should be the same, within experimental error.

2. Experimentally, measuring the increase in the pressure of the CO_2 gas with respect to time for several concentrations would provide information to determine the reaction order.

3. The initial rate provides reproducible information because competing reactions are not significant at this stage. With time the rate of a reaction decreases so the initial rate is what is measured.

4. Activation energy and temperature are explicitly in the equation because these factors influence the energy of the collision. The Arrhenius constant, A, includes the frequency of collisions with correct geometry necessary for reaction.

5. The mechanism would follow the given rate law with N_2O_2 as a reactive intermediate if Step 1 is slow. Then the overall reaction is $2 NO(g) + H_2(g) \longrightarrow N_2O(g) + H_2O(g)$.

6. The initial rate of reaction is determined for several difference concentrations of reactant. From this information the rate law may be determined and the order of the reaction with respect to each reactant. Once this information is known, the rate constant, k, may be calculate. It will be a constant for each reaction, within experimental error.

7. Rate $= k [NO]^2[F_2]^1$ From experiments 1–3, a value of k= 1.0×10^2 $L^2/mol^2 \cdot s$ is determined. This is used with the concentrations of Experiment 4 for find a rate of 25 mol/L·s.

8. A catalyst changes the path of the reaction by lowering the activation energy which makes it easier for the reaction to take place. This contrasts with a change in temperature which does not change the path of the reaction, but rather increases the energy of the particles making it more probable for a reaction to occur.

9. Kinetic data allows one to determine the order of the reaction based on observed rates. Mechanisms are an attempt to explain the observed data by various models that are "reasonable" in accord with other principles of chemistry. Mechanisms may be disproved when an alternate path is proposed which still follows the same kinetics. Since we cannot (at this time) actually see what the particles are doing, we must rely on deductive reasoning. Sometimes we are wrong in our thought processes, because several explanations can account for the observations.

10. In order to rid exhaust from automobiles of harmful gases such as NO, CO, and unreacted hydrocarbons, a catalytic converter is used to promote the reactions with oxygen to form less harmful, N_2, CO_2, and H_2O. Typical equations are

$2 CO + O_2$ --catalyst---> CO_2
$2 NO$ --catalyst--> $N_2 + O_2$
$C_xH_y + O_2$ --catalyst---> $CO_2 + H_2O$

Section B: Multiple Choice

11. For the reaction   A ` 2B + C the rate could be expressed as $-\Delta[A]/\Delta t$. An equivalent expression is

   a) $-\dfrac{1}{2}\dfrac{\Delta[B]^2}{\Delta t}$ b) $-\dfrac{1}{2}\dfrac{\Delta[B]}{\Delta t}$ c) $+\dfrac{1}{2}\dfrac{\Delta[B]}{\Delta t}$ d) $+\dfrac{\Delta[B]^2}{\Delta t}$ e) $+\dfrac{\Delta[B][C]}{\Delta t}$

12. For the gas phase reaction $3 H_2 + N_2 \longrightarrow 2 NH_3$, how does the rate of disappearance of H_2 compare to the rate of production of NH_3?
   a) the initial rates are equal
   b) the rate of disappearance of H_2 is 1/2 the rate of appearance of NH_3
   c) the rate of disappearance of H_2 is 3/2 the rate of appearance of NH_3
   d) the rate of disappearance of H_2 is 2/3 the rate of appearance of NH_3
   e) the rate of disappearance of H_2 is 1/3 the rate of appearance of NH_3

13. Consider the reaction $S_2O_8{}^{2-} + 3 I^- \longrightarrow 2 SO_4{}^{2-} + I_3^-$
   Which one of the following rate expressions would give the same value as the rate of disappearance of $S_2O_8{}^{2-}$?
   a) rate $= -3 \Delta[I^-]/\Delta t$ b) rate $= -1/3\,(\Delta[I^-])/\Delta t$
   c) rate $= -2\,(\Delta[SO_4{}^{2-}])/\Delta t$ d) rate $= -\Delta[I_3^-]/\Delta t$
   d) rate $= -1/2\,(\Delta[SO_4{}^{2-}])/\Delta t$

14. For the reaction $Cl_2 + 2 NO \longrightarrow 2 NOCl$
   which one of the following rate expressions would give the same value for the rate as the change in molarity of Cl_2?
   a) $-2\Delta[NO]/\Delta t$ b) $+1/2\,(\Delta[NOCl])/\Delta t$
   c) $-2\,(\Delta[NOCl]/\Delta t)$ d) $-1/2\,(\Delta[NOCl]/\Delta t)$
   e) $+1/2\,(\Delta[NO]/\Delta t)$

15. For the reaction

$$C_2H_2 \text{ (g)} + 3 N_2O_{(g)} \longrightarrow 3 N_{2(g)} + 2 CO_{(g)} + H_2O_{(g)}$$

The rate of disappearance of C_2H_2 was measured as -1.5×10^{-5} M/s. Of the following experimental measurements which one verifies the result of the C_2H_2 rate determination?

a) $\Delta[H_2O]) / \Delta t = -3.0 \times 10^{-5}$ M/s   b) $\Delta[N_2O] / \Delta t = -4.5 \times 10^{-5}$ M/s

c) $\Delta[N_2] / \Delta t = -0.50 \times 10^{-5}$ M/s   d) $\Delta[CO] / \Delta t = +1.5 \times 10^{-5}$ M/s

e) $\Delta[N_2] / \Delta t = +1.5 \times 10^{-5}$ M/s

16. For the reaction below, which equation represents the rate of reaction?

$$6 CH_2O + 4 NH_3 \longrightarrow (CH_2)_6N_4 + 6 H_2O$$

a) rate $= (1/6) \Delta[CH_2O] / \Delta t$ b) rate $= -4 \Delta[NH_3] / \Delta t$

c) rate $= -\Delta[(CH_2)_6N_4] / \Delta t$ d) rate $= 6 \Delta[H_2O] / \Delta t$

e) rate $= (1/6) \Delta[H_2O] / \Delta t$

17. A student analyzed a second order reaction and obtained the graph at the right but forgot to label the axes. What should the labels be for the X and Y coordinates respectively ?

a) time, ln [A]

b) time, [A]

c) temperature, [A]

d) temperature, ln[A]

e) time, 1/[A]

18. For the gaseous reaction $2 A \longrightarrow B$ the results of a plot of ln[A] versus time is in the figure. What conclusion regarding the reaction order may be made?

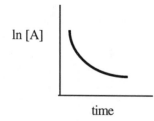

ln [A]

time

a) the reaction is first order
b) the reaction is not first order
c) the reaction is second order
d) the reaction is not second order
e) the reaction is third order

19. For a first order reaction, which of the following can be plotted versus time to give a straight line.

   a) ln [A]   b) ln k   c) ln [1/A]

   d) 1/[A]   e) [A]

20. For the reaction   2 A ⟶ B   the reaction order is

   a) zero order

   b) first order

   c) second order

   d) impossible to predict without experimental rates at various concentrations of A

   e) impossible to predict without knowing the heat of reaction

21. For the reaction   2 A + 2 B ⟶ 3 C, it was determined that the reaction is third order overall. The rate law for this reaction might be

   a) rate = k $[A]^2[B]^2$ b) rate = k $[A][B]^2$ c) rate = k $[C]^3$

   d) rate = k [A][B]   e) rate = k $[A][B]^3$

22. In studying the reaction A ⟶ B, a plot of 1/[A] vs. time provided a straight line with a slope of 1.22 /M· s. What is the rate law expression for the reaction?

   a) k = –1.22   b) $\Delta[A]/\Delta t = -1.22[A]^2$

   c) $\Delta[A]/\Delta t = +1.22[A]^2$ d) $\Delta[A]/\Delta t = -1.22[A]$

   e) $\Delta[A]/\Delta t = -0.829[A]$

23. In studying the reaction 2 X ⟶ Z, a plot of ln[X] vs. time provided a straight line with a slope of –1.34 /s. What is the rate law expression for the reaction?

   a) k = +1.34   b) $\Delta[X]/\Delta t = -1.34[X]^2$

   c) $\Delta[X]/\Delta t = +1.34[X]^2$ d) $\Delta[X]/\Delta t = -1.34[2X]^2$

   e) $\Delta[X]/\Delta t = -1.34[X]$

24. What are the units for k, the rate constant, in a first order reaction where the time unit is seconds (s)?

   a) mol / L· s   b) mol/L   c) 1/s

   d) s· mol / L   e) s· L / mol

25. For a reaction, the rate law is rate = $k[A]^1[B]^0[C]^1$. What are the units for k where the time unit is seconds (s)?

    a) (mol / L· s)

    b) L/mol· s

    c) L^2/mol^2· s

    d) mol^2 / L^2 · s

    e) mol^3 / L^3 · s

26. What is the rate law for the reaction A + B $\longrightarrow$ 2 C based on the following kinetic data?

Experiment Number	Initial Conc. (mol/L) $[A]_o$	Initial Conc. (mol/L) $[B]_o$	Initial rate of reaction (mol/L· s)
1	0.40	0.10	3.5 x 10^3
2	0.20	0.10	1.8 x 10^3
3	0.20	0.50	4.5 x 10^4

    a) rate = k $[A]^1[B]^2$

    b) rate = k $[A]^{1/2}[B]^5$

    c) rate = k $[A]^2[B]^1$

    d) rate = k $[A]^1[B]^{1/5}$

    e) rate = k $[A]^{1/2}[B]^2$

27. Kinetic data for the following reaction was determined experimentally:

$$4 NO_{2(g)} \quad + \quad O_{2(g)} \longrightarrow 2 N_2O_{5\,(g)}$$

Experiment Number	Initial Conc. (mol/L) $[NO_2]_o$	Initial Conc. (mol/L) $[O_2]_o$	Initial rate of reaction (mol/ L·s)
1 0.40		0.10	3.3
2 0.20		0.10	1.7
3 0.20		0.50	41.

What is the rate law for the reaction?

    a) rate = k $[NO_2]^4[O_2]^1$

    b) rate = k $[NO_2]^{1/2}[O_2]^2$

    c) rate = k $[NO_2]^2[O_2]^2$

    d) rate = k $[NO_2]^1[O_2]^1$

    e) rate = k $[NO_2]^1[O_2]^2$

28. Kinetic data for the following reaction was determined experimentally:

$$2 X \quad + \quad Y \longrightarrow 3 Z$$

Experiment Number	Initial Conc. (mol/L) $[X]_o$	Initial Conc. (mol/L) $[Y]_o$	Initial rate of reaction (mol/L·s)
1	0.15	0.10	6.4 x 10^2
2	0.15	0.20	2.6 x 10^3
3	0.30	0.30	1.2 x 10^4

What is the rate law for the reaction?

    a) rate = k $[X]^2[Y]^1$

    b) rate = k $[X]^1[Y]^3$

    c) rate = k $[2X]^1[Y]^3$

    d) rate = k $[X]^1[Y]^2$

    e) rate = k $[X]^2[Y]^3$

29. Kinetic data for the following reaction was determined experimentally:

$$2 NO_{(g)} + Cl_{2(g)} \longrightarrow 2 NOCl_{(g)}$$

Experiment Number	Initial Conc. (mol/L) $[NO]_o$	Initial Conc. (mol/L) $[Cl_2]_o$	Initial rate of reaction (mol/L·s)
1	0.20	0.10	0.63
2	0.20	0.30	5.70
3	0.80	0.10	2.58
4	0.40	0.20	?

Predict an observed rate of reaction for Experiment 4.

a) 0.5 mol/L·s   b) 2.5 mol/L·s   c) 4.2 mol/L·s

d) 5.1 mol/L·s   e) 10.1 mol/L·s

30. Kinetic data for the following reaction was determined.

$$A + 2 B(g) \longrightarrow 2 C(g)$$

Experiment Number	Initial Conc. (mol/L) $[A]_o$	Initial Conc. (mol/L) $[B]_o$	Initial rate of disappearance of A (mol/L·s)
1	0.15	0.10	4.10×10^3
2	0.45	0.10	3.69×10^4
3	0.15	0.30	1.20×10^4

The rate expression for the above reaction is

a) rate = k $[A]^2[B]$ b) rate = k $[A]^2[2 B]^2$ c) rate = k $[A][B]^2$

d) rate = k $[A][2 B]$ e) rate = k $[A][2 B]^2$

31. Kinetic data for the following reaction was determined experimentally:

$$4 NO_{2(g)} + O_{2(g)} \longrightarrow 2 N_2O_{5 (g)}$$

Experiment Number	Initial Conc. (mol/L) $[NO_2]_o$	Initial Conc. (mol/L) $[O_2]_o$	Initial rate of reaction (M/sec)
1	0.40	0.10	3.1
2	0.20	0.10	0.78
3	0.10	0.40	3.1

What is the rate law expression for the reaction?

a) k=$[NO_2]^2[O_2]^1$ b) k= $[NO_2]^1[O_2]^2$ c) rate = k $[NO_2]^2[O_2]^1$

d) rate = k $[NO_2]^1[O_2]^2$ e) rate = k $[NO_2]^2[O_2]^2$

32. The reaction X $\longrightarrow$ Y follows first order kinetics with k = 0.83 /min. If the initial
    concentration of X is 3.6 M, what is the concentration of X after 15 minutes?
    a) 0.046 M   b) 0.230 M   c) 1.1×10^{-1} M
    d) 1.84×10^{-3} M   e) 1.4×10^{-5}M

33. The reaction A ` B follows first order kinetics with k = 0.16/min. If the initial
    concentration of A is 2.4 M, what is the concentration of B after 5.0 minutes?
    a) 1.3 M   b) 1.1 M   c) 0.83 M
    d) 1.6 M   e) 3.4 M

34. Which of the following rate expressions is bimolecular overall?
    a) rate = k [A]   b) rate = k [A][B]   c) rate = k $[A]^2[B]^2$
    d) rate = k $[A]^2[B]$ e) rate = k [2A]

35. What is unique about the half-life of any first order reaction at 25°C?
    a) The units are always sec^{-1}
    b) The value depends only on the rate constant k.
    c) The value depends only on the initial concentration of reactant.
    d) $\Delta[A_0]/\Delta t = 1$
    e) $\Delta[A_0]/\Delta t = 1/2$

36. After five half-life periods for a first order reaction, what is the molarity of a reagent
    initially at 0.366 M?
    a) 1.14×10^{-2} b) 3.12×10^{-2} c) 6.57×10^{-3}
    d) 3.12×10^{3} e) 7.32×10^{-2}

37. If the half-life of a first order process is 3.00 minutes, the rate constant for the process is
    a) 1.50 /min   b) 1.05 /min   c) 4.34/min
    d) 0.405 /min   e) 0.231 /min

38. The decomposition of HCO_2H follows first-order kinetics:

$$HCO_2H_{(g)} \longrightarrow CO_{2(g)} + H_{2(g)}$$

The half-life for the reaction at 550°C is 24 seconds. How many seconds are needed for formic acid, initially 0.82 M, to decrease to 0.018M?

a) 1.3×10^2 seconds   b) 1.1×10^3 seconds   c) 2.9×10^3 seconds

d) 7.4×10^3 seconds   e) 9.0×10^4 seconds

39. The half-life for a first order reaction at 320°C is 11.6 seconds. How many seconds are needed for the reactant, initially at 0.720 M, to decrease to 0.0220M?

a) 28.4 seconds   b) 36.9 seconds   c) 43.6 seconds

d) 58.5 seconds   e) 77.1 seconds

40. The half-life for a first order reaction at 550°C is 85 seconds. How long would it take for 23% of the reactant to decompose?

a) 0.82 seconds   b) 26.1 seconds   c) 32 seconds

d) 43.6 seconds   e) 180 seconds

41. The decomposition of phosphine, PH_3, follows first-order kinetics:

$$4 PH_{3(g)} \longrightarrow P_{4(g)} + 6 H_{2(g)}$$

The half-life for the reaction at 550°C is 81.3 seconds. How long does it take for the reaction to be 78.5% complete?

a) 8.52 seconds   b) 28.4 seconds   c) 63.8 seconds

d) 117 seconds   e) 180 seconds

42. What is the half-life of a first order reaction if it take 143 seconds for the concentration to decrease from 1.50 M to 0.0415 M?

a) 0.0251 seconds   b) 3.58 seconds   c) 4.96 seconds

d) 27.6 seconds   e) 97.2 seconds

43. What is the half-life of a first order reaction which is 15% complete after 210 seconds?

a) 7.74 seconds   b) 32 seconds   c) 76.7 seconds

d) 178 seconds   e) 896 seconds

44. The decomposition of N_2O_5 follows first-order kinetics:

$$2\ N_2O_{5(g)} \longrightarrow 4\ NO_{2(g)} + O_{2(g)}$$

The half-life for the reaction is 48.6 seconds. If the initial pressure of N_2O_5 is 185 torr, what will be its pressure after 275 seconds?

a) 3.72 torr   b) 9.14 torr   c) 12.5 torr

d) 37.2 torr   e) 142 torr

45.   If the half life of a first order reaction is 12.0 days, how long does it take for 60.0% of the starting material to disappear?

a) 7.2 days   b) 15.9 days   c) 16.8 days

d) `19.2 days   e) 22.3 days

46. The Arrhenius equation, $k = Ae^{-Ea/RT}$, may be used to calculate the activation energy from the slope of a line plotted with what parameters ?

a) ln k vs 1/Temperature   b) ln k vs 1/time   c) 1/k vs Temperature

d) 1/k vs 1/time   e) ln k vs e^{-T}

47. Which of the following reactions will have the greatest rate at 298 K? Assume that the frequency factor A is the same for all reactions.

a) $\Delta E = +10$ kJ/mol $E_a = 25$ kJ/mol   b) $\Delta E = -10$ kJ/mol $E_a = 25$ kJ/mol

c) $\Delta E = +10$ kJ/mol $E_a = 10$ kJ/mol   d) $\Delta E = -10$ kJ/mol $E_a = 50$ kJ/mol

e) $\Delta E = -10$ kJ/mol $E_a = 15$ kJ/mol

48. Adding a catalyst will NOT change the magnitude of which energy change?

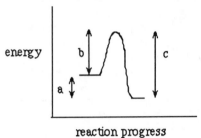

a) a only   b) b only   c) c only   d) both a and b   e) both b and c

49. In general, as the temperature increases, the rate of a chemical reaction
    a) increases due to an increased activation energy.
    b) increases only for an endothermic reaction.
    c) increases due to a greater number of effective collisions.
    d) increases because bonds are weakened.
    e) is not changed.

50. Which of the following reactions will have the greatest rate at 298 K? Assume that the frequency factor A is the same for all reactions.

    a) $\Delta E = +10$ kJ/mol $E_a = 25$ kJ/mol   b) $\Delta E = -10$ kJ/mol $E_a = 25$ kJ/mol
    c) $\Delta E = +10$ kJ/mol $E_a = 11$ kJ/mol   d) $\Delta E = -10$ kJ/mol $E_a = 50$ kJ/mol
    e) d) $\Delta E = +25$ kJ/mol $E_a = 25$ kJ/mol

51. A reaction has an activation energy of 40 kJ and an overall energy change of –100 kJ. What is the potential energy diagram which best describes this reaction?

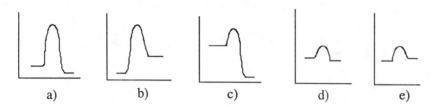

a)   b)   c)   d)   e)

52. If the activation energy for the forward reaction of a given process is +110 kJ and the activation energy for the reverse reaction of the same process is +60.0 kJ, then the energy change for the overall process is

a) –50kJ   b) +50 kJ   c) –170 kJ   d) + 170 kJ   e) -60 kJ

53. In basic solution, $(CH_3)_3CCl$ reacts according to the equation

$$(CH_3)_3CCl \ + \ OH^- \longrightarrow \ (CH_3)_3COH \ + \ Cl^-$$

The accepted mechanism for the reaction is

$(CH_3)_3CCl \longrightarrow (CH_3)_3C^+ \ + \ Cl^-$ (slow)

$(CH_3)_3C^+ \ + \ OH^- \longrightarrow \ (CH_3)_3COH$ (fast)

What is the rate law expression for the reaction?

a) rate = k $[(CH_3)_3C^+]^2[OH^-]$ b) rate = k $[(CH_3)_3C^+][OH^-]^2$ c) rate = k $[Cl^-]$

d) rate = k $[(CH_3)_3CCl]$ d) rate = k $[(CH_3)_3CCl][OH^-]$

Chapter 15: Multiple Choice Answers

11. c	21. b	31. d
12. c	22. b	32.
13. b	23. e	33. a
14. b	24. c	34. b
15. b	25. b	35. b
16. e	26. a	36. a
17. e	27. e	37. c
18. b	28. d	38. a
19. a	29. d	39. d
20. d	30. a	40. c

41. e	51. c
42. d	52. b
43. e	53. d
44. a	
45. b	
46. a	
47. c	
48. a	
49. c	
50. c	

Chapter 16
Chemical Equilibria

Section A: Free Response

1. Consider the reaction: $2 A(g) + B(g) \rightleftharpoons 2 C (g)$

    Experimentally how could you determine if the reaction is a equilibrium. Design two experiments which utilize common measurements and predict the outcome of each.

2. What is meant by the term "dynamic equilibrium'?

3. Your job is to increase the yield of the industrial process for the production of ammonia according to the following equation:

    $N_2(g) + 3 H_2(g) \rightleftharpoons 2 NH_3(g)$ $\Delta H = -46.1 kJ$ at 25°C

    How could the pressure and temperature be changed to increase the yield of NH_3?

4. Consider the gaseous equilibrium process for compound A which is colorless, and compound B which is blue, according to the equation:

    $2 A (g) \rightleftharpoons B (g)$
    colorless   blue

    The size of the container is suddenly halved and a new equilibrium is established. Use LeChatelier's principle to explain
    a) the change in color (if any) that would be observed initially when the size of the container is decreased.
    b) the change in color (if any) that would be observed in the smaller container after time has passed.

5. Consider the reaction of carbon disulfide and chlorine according to the following equation:

    $CS_2(g) + 3 Cl_2(g) \rightleftharpoons S_2Cl_2(g) + CCl_4(g)$

    When 1.2 moles of CS_2 and 3.6 moles of Cl_2 are placed in a 1.00 liter container and allowed to reach equilibrium, the mixture contains 0.90 mol CCl_4. Calculate the equilibrium concentrations of all species and K_c.

6. At a particular temperature a 1.0 liter flask contained an equilibrium mixture of the gases in the reaction below.

$$CO \text{ (g)} + H_2O\text{(g)} \rightleftharpoons H_2\text{(g)} + CO_2\text{(g)}$$

   a) If the concentrations are 0.10 M CO, 0.10 M H_2O, 0.10 M H_2, and 0.40 M CO_2, calculate K_c.

   b) To the equilibrium mixture an additional 0.30 mol H_2 was added. Calculate the new equilibrium concentration of each species. Assume no change in temperature.

7. At a particular temperature $K_c = 3.6$ for the following reaction. If a 2.00 liter flask is filled with 1.5 mole SO_3, 2.5 moles SO_2 and 0.5 mole O_2, will the concentration of O_2 increase, decrease, or remain the same as equilibrium is established? Explain.

$$2 SO_{3(g)} \rightleftharpoons 2 SO_{2(g)} + O_{2(g)} \qquad K_c = 3.6 \text{ M}$$

8. A mixture of 4.0×10^{-3} moles of H_2 and 6.0×10^{-3} moles of I_2 is placed in a 4.0 liter container at $448°C$ and allowed to reach equilibrium. Analysis of the equilibrium mixture shows that 8.0×10^{-4} moles of HI have been formed. Calculate K_c at $448°C$ for the reaction

$$H_{2(g)} + I_{2(g)} \rightleftharpoons 2 HI_{(g)}$$

9. The reaction $A + 2 B \text{ ---> } 2 C$ proceeds by the following mechanism:
   Elementary Step 1: $A + B \rightleftharpoons D$ \qquad a fast equilibrium
   Elementary Step 2: $D + B \text{ ---> } 2 C$ \qquad a slow step

   Derive the rate law for this reaction.

10. A flask contains the following system at equilibrium:

$$Ca(OH)_2\text{(s)} \rightleftharpoons Ca^{2+}\text{(aq)} + 2 OH^- \text{ (aq)}$$

   Use LeChatelier's Principle to discuss ways the solubility of $Ca(OH)_2$ may be increased.

Key Concepts for Free Response

1. Experiment #1: Measure the concentration of one species at several points in time. If the concentration does not change the system is a equilibrium. Experiment #2: Measure the concentrations of all species and calculate the ratio $C^2/[A]^2[B]$. Then add more of one reagent. If the system is at equilibrium the concentrations of all the other species will change so that the ratio of products/reactants remains the same.
      Experiment #3: Measure the total pressure in the flask. Measure the pressure of C. Then you can calculate the pressures of A and B. Calculate Kp. Keep the temperature constant. Change the volume. readjust to reach a new equilibrium

2. The rates of the forward and reverse reactions are the same. The concentrations of all reactant and products remains the same which is called the "steady state."
   The ratio of the molar concentrations of the products to reactants is the same.

3. The reaction is exothermic so a low temperature will shift the equilibrium to favor NH_3. Since there are fewer particles on the product side, a higher pressure will also increase the formation of NH_3.

4. The original color is a light blue because it is a mixture of colorless and blue compounds. When the size of the container is decreased, the concentration of all species is increased and it appears darker blue. In time the increased pressure within the vessel will favor the forward reaction, so the color becomes darker and darker blue.

5.

	$[CS_2]$	$[Cl_2]$	$[S_2Cl_2]$	$[CCl_4]$
[Initial]	1.2	3.6	0	0
change	-x	-3x	+x	+x
[At Equilibrium]	1.2 - 0.9 = 0.3	3.6 - 2.7 = 0.9	0.9	0.9

   K_c = 3.7 or 4   (1 sig fig allowed)

6. K_c = 4;   [CO] = 1.66; [H_2O] = 1.66; [H_2] = 3.33; [CO_2] = 3.33

7. The reaction quotient, Q, is 0.60 for the given molar quantities. Since Q<K, the reaction will shift toward more products and the concentration of O2 will increase as equilibrium is established.

8. K_c = 8.9 x 10^{-2}

9. The rate depends on the slow step:
   [D] can be calculated from the equilibrium step

   Rate = k[D][B]
   K_c = [D] / [A][B]
   [D]= K_c[A][B]
   Rate = k [A][B]^2

   Substituting and combining constants

10. Shifting the reaction toward the products would increase the solubility of the $Ca(OH)_2$. Removing the OH⁻ by adding an acid would be an effective method.

Section B: Multiple Choice

11. At equilibrium, what is equal?

a) concentrations of products and reactants

b) rate constants for the forward and reserve reactions

c) the rate of the forward and reverse reaction

d) the partial pressures of the reactants and products

e) the reaction quotient and the rate of both reactions

12. In which case does the reaction go farthest to completion?

a) K= 1000   b) K = 1   c) K = 10^4

d) K= 0.001   e) K= 10^{-5}

13. Under which of the following conditions does the equilibrium constant K change for the reaction

$$H_2(g) + I_2(g) \rightleftharpoons 2\,HI\,(g)$$

a) changing the size of the container

b) introducing more I_2 into the container

c) measuring the molar concentrations instead of pressures

d) changing the temperature

e) none of these, it is always constant

14. What is the expression for K_c for the following reaction?

$$2\,NO_2(g) \rightleftharpoons 2\,NO(g) + O_2(g)$$

a) $K_c = [NO]\,[O_2]\,/\,[NO_2]^2$ b) $K_c = [\,2\,NO]^2\,[O_2]\,/\,[\,2\,NO_2]^2$

c) $K_c = [NO]^2\,[O_2]\,/\,[NO_2]^2$ d) $K_c = [NO_2]^2/\,[NO]^2\,[O_2]^2$

e) $K_c = [NO]^2 + [O_2]\,/\,[NO_2]^2$

15. What is the equilibrium expression, K_c, for the following reaction?

$$Ca_3(PO_4)_{2(s)} \rightleftharpoons 3\,Ca^{2+}{}_{(aq)} + 2\,PO_4{}^{3-}{}_{(aq)}$$

a) $K_c = [Ca_3(PO_4)_2]\,/[Ca^{2+}][PO_4{}^{3-}]^2$ b) $K_c = [Ca^{2+}][PO_4{}^{3-}]$

c) $K_c = [Ca^{2+}]^3[PO_4{}^{3-}]^2/[Ca_3(PO_4)_2]$ d) $K_c = [Ca^{2+}]^3[PO_4{}^{3-}]^2$

e) $K_c = [3\,Ca^{2+}]^3[2\,PO_4{}^{3-}]^2/[Ca_3(PO_4)_2]$

16. For the reaction below, what is the expression for K_C?

$$2 H_2(g) + 2 FeO(s) \rightleftharpoons 2 Fe(s) + 2 H_2O(g)$$

a) $K_C = [Fe]^2 [H_2O]^2 / [H_2]^2 [FeO]^2$

b) $K_C = [H_2O]^2 / [H_2]^2 [FeO]^2$

c) $K_C = [Fe]^2 / [H_2]^2 [FeO]^2$

d) $K_C = [H_2O]^2 / [H_2]^2$

e) $K_C = [2 Fe]^2 [2 H_2O]^2 / [2 H_2]^2 [2 FeO]^2$

17. For the reaction below, what is the equilibrium constant ?

$$2 H_2O_2(aq) \rightleftharpoons O_2(g) + 2 H_2O(l)$$

a) $K_c = [O_2(g)]$

b) $K_c = [O_2(g)][H_2O(l)]^2 / [H_2O_2(aq)]^2$

c) $K_c = [O_2(g)] / [H_2O_2(aq)]^2$

d) $K_c = [O_2]^{1/2} / [H_2O_2(aq)]^2$

e) $K_c = [O_2(g)] / [2 H_2O_2(aq)]^2$

18. For the reaction below, what is the expression for K_p ?

$$SF_6(g) \rightleftharpoons S(s) + 3 F_2(g)$$

a) $P_{SF_6} / (P_S \cdot P_{F_2}^3)$

b) $(P_S \cdot P_{F_2}^3) / P_{SF_6}$

c) $P_{F_2}^3 / P_{SF_6}$

d) $(P_S \cdot 3P_{F_2})^3 / P_{SF_6}$

e) $P_{SF_6} / P_{F_2}^3$

19. For the reaction $2 A + B_2 \rightleftharpoons C$, $K_C = 1.2$. What is K_C for the reaction

$6 A + 3 B_2 \rightleftharpoons 3 C$?

a) 1.2   b) 3.6   c) 1.7   d) 7.2   e) 1.1

20. For the reaction $2 A_2 + B \rightleftharpoons 3 C$, $K_C = 5.8$. What is K_C for the reaction

$3 C \rightleftharpoons 2 A_2 + B$?

a) 0.17   b) 2.4   c) 2.9   d) 3.6   e) 17.4

21. If $K_C = 0.44$ for the reaction $2 NOBr(g) \rightleftharpoons 2 NO(g) + Br_2(g)$ at a particular temperature, what is K_C for the following reaction?

$$NOBr(g) \rightleftharpoons NO(g) + 1/2 Br_2(g)$$

a) 0.44   b) 0.66   c) 0.22   d) 0.19   e) 2.3

22. If $K_c = 1.6 \times 10^{-10}$ àt 300ºC for the reaction $2 SO_{3(g)} \rightleftharpoons 2 SO_{2(g)} + O_{2(g)}$

   then what is K_c at 300ºC for the reaction below?
$$SO_2(g) + 1/2\, O_2(g) \rightleftharpoons SO_3(g)$$

   a) 1.3×10^{-5} b) 6.25×10^9 c) 7.9×10^4
   d) 1.6×10^{-10} e) 0.80×10^{-15}

23. If $K_c = 2.5 \times 10^4$ at a particular temperature for the reaction
$$2 H_{2(g)} + O_{2(g)} \rightleftharpoons 2 H_2O(g)$$

   then what is K_c at the same temperature for the following reaction?
$$H_2O(g) \rightleftharpoons H_{2(g)} + 1/2\, O_{2(g)}$$

   a) 6.3×10^{-3} b) 1.2×10^4 c) 1.4×10^8
   d) 1.6×10^2 e) 6.3×10^8

24. The following two-step process has equilibrium constants K_1 and K_2.
   Step 1: $H_{2\,(g)} + ICl_{(g)} \rightleftharpoons HI_{(g)} + HCl_{(g)}$ K_1
   Step 2: $HI_{(g)} + ICl_{(g)} \rightleftharpoons HCl_{(g)} + I_{2(g)}$ K_2

   Overall: $H_{2\,(g)} + 2 ICl_{(g)} \rightleftharpoons 2 HCl_{(g)} + I_{2(g)}$ K_3

   What is the expression for the equilibrium constant for the overall reaction, K_3?
   a) $K_3 = K_1 K_2$ b) $K_3 = 1/ [K_1]^2$
   c) $K_3 = K_1 + K_2$ d) $K_3 = (K_1 K_2)^2$
   e) $K_3 = K_2 / K_1$

25. Consider the reactions $2 SO_{2\,(g)} + O_{2\,(g)} \rightleftharpoons 2 SO_{3\,(g)}$ K_1
             and $SO_{3\,(g)} \rightleftharpoons SO_{2\,(g)} + 1/2\, O_{2\,(g)}$ K_2

   What is the relationship between the K values of the two reactions?
   a) $K_2 = K_1^{1/2}$ b) $K_2 = K_1^{-1/2}$ c) $K_2 = K_1^2$
   d) $K_2 = -(K_1)^2$ e) $K_2 = 1/2 (K_1)^2$

26. What is the relationship between K_p and K_c for the following reaction?
$$CH_3OH(g) \rightleftharpoons CO(g) + 2 H_2(g)$$

   a) $K_p = K_c$ b) $K_p = K_c(RT)^{-1}$ c) $K_p = K_c(RT)$
   d) $K_p = K_c(RT)^{-2}$ e) $K_p = K_c(RT)^2$

27. What is the relationship between K_p and K_c for the following reaction?
$$2\,SO_{2\,(g)} + O_{2\,(g)} \rightleftharpoons 2\,SO_{3\,(g)}$$

a) $K_p = K_c$ 　　　　b) $K_p = K_c(RT)^{-1}$ 　　　　c) $K_p = K_c(RT)^{1}$

d) $K_p = K_c(RT)^2$ 　　　　e) $K_p = K_c(RT)^{-2}$

28. What is the relationship between K_p and K_c for the equation below?
$$CO(g) + Cl_2(g) \rightleftharpoons COCl_2(g)$$

a) $K_p = K_c$ 　　　　b) $K_p = K_c(RT)^{-1}$ 　　　　c) $K_p = K_c(RT)^{1}$

d) $K_p = K_c(RT)^{-2}$ 　　　　e) $K_p = K_c(RT)^{2}$

29. In which of the following equations is $K_c = K_p$?

a) $SO_2Cl_2(g) \rightleftharpoons SO_2(g) + Cl_2(g)$

b) $C(s) + H_2O(g) \rightleftharpoons CO(g) + H_2(g)$

c) $2\,SO_3(g) \rightleftharpoons 2\,SO_2(g) + O_2(g)$

d) $2\,HI(g) \rightleftharpoons H_2(g) + I_2(g)$

e) $NO + 1/2\,Br_2(g) \rightleftharpoons NOBr(g)$

30. A gaseous mixture at 300°C is at equilibrium. It contains 8.7 mol HCl, 0.21 mol H_2 and 0.43 mol Cl_2 in a 1.00 L flask. Calculate K_c for the reaction
$$2\,HCl(g) \rightleftharpoons H_2(g) + Cl_2(g)$$

a) 96 　　b) 1.0×10^{-2} 　　c) 1.2×10^{-3} 　　d) 8.4×10^2 　　e) 1.1×10^{-4}

31. A gaseous mixture at 500°C is at equilibrium according to the following equation:
$$2\,NOBr(g) \rightleftharpoons 2\,NO(g) + Br_2(g)$$

It contains 4.0 mol NOBr, 0.50 mol NO and 0..25 mol Br_2 in a 1.00 L flask. Calculate K_c.

a) 0.31 　　b) 3.9×10^{-3} 　　c) 9.7×10^{-2} 　　d) 4.1×10^3 　　e) 2.4×10^{-4}

32. Consider an equilibrium mixture of oxygen and ozone according to the equation
$$3\,O_2(g) \rightleftharpoons 2\,O_3(g)$$
The partial pressure of O_2 was measured in a flask at equilibrium as 1.25 atm and the total pressure in the flask was 1.75 atm. Calculate K_p. Constant temperature was maintained.

a) 8.0 x 10^{-3} b) 0.90   c) 0.13   d) 1.6   e) 2.7

33.   A chemist prepared a sealed tube with 0.96 atm of CO and 1.02 atm of Cl_2 at 500K. The pressure dropped as the following reaction occurred. When equilibrium was achieved, the pressure in the tube dropped from an initial 2.05 atm to 1.45 atm. Calculate K_p.
$$CO(g) + Cl_2(g) \rightleftharpoons COCl_2 (g)$$

a) 1.7   b) 2.0   c) 3.1   d) 0.60   e) 0.89

34. A chemist prepared a sealed tube with 0.85 atm of PCl_5 at 500K. The pressure increased as the following reaction occurred. When equilibrium was achieved, the pressure in the tube had increased to 1.25 atm. Calculate Kp.
$$PCl_5(g) \rightleftharpoons PCl_3(g) + Cl_2 (g)$$

a) 0.36   b) 0.19   c) 0.10   d) 0.047   e) 0.089

35.   A 1.00 liter flask contained 0.24 mol NO_2 at 700 K. which decomposed according to the following equation. When equilibrium was achieved, 0.14 mol NO was present. Calculate Kc.
$$2\,NO_2(g) \rightleftharpoons 2\,NO(g) + O_2(g)$$

a) 0.098   b) 0.14   c) 1.1 x 10^{-2} d) 5.7 x 10^3 e) 9.6 x 10^{-3}

36. Ammonia decomposes according to the following equation.
$$2\,NH_3(g) \rightleftharpoons N_2(g) + 3\,H_2(g)$$
When 0.60 mol NH_3 is introduced into a 1.00 liter flask at 850 K, the equilibrium concentration of NH_3 is measured as 0.12 M. Calculate K_c.

a) 0.031   b) 0.024   c) 0.090   d) 0.144   e) 0.78

37. Calculate the equilibrium constant, K_p, for the reaction
$$3\,A(g) \;\rightleftharpoons\; B(g) + 2\,C(g)$$

given that a 1.0 L vessel was initially filled with 5.0 atm of pure A and after equilibrium was reached, the partial pressure of gas A was 3.5 atm.

a) 0.14   b) 4.9   c) 9.8   d) 1.2×10^{-2} e) 2.1×10^{-3}

38. Exactly 2.00 moles of SO_2 and 4.00 moles of O_2 were introduced into a 1.00 liter flask at 450°C. When equilibrium was achieved the number of moles of O_2 was 3.14. Calculate K.
$$2\,SO_2(g) + O_2(g) \;\rightleftharpoons\; 2\,SO_3(g)$$

a) 0.109   b) 1.16   c) 1.96   d) 11.2   e) 12.0

39. A mixture of 5.00×10^{-3} moles of H_2 and 2.00×10^{-3} moles of I_2 is placed in a 5.00 liter container at 448°C and allowed to reach equilibrium. Analysis of the equilibrium mixture shows that 1.86×10^{-3} moles of HI have been formed. Calculate K_c at 448°C for the reaction
$$H_2(g) + I_2(g) \;\rightleftharpoons\; 2\,HI(g)$$

a) 0.79   b) 15.0   c) 4.2×10^3 d) 2.2×10^5 e) 9.3×10^{-5}

40. A mixture of 0.30 mol NO and 0.30 mole CO_2 is placed in a 2.00 L flask and allowed to reach equilibrium. Analysis of the equilibrium mixture indicated that 0.10 mol of CO was present. Calculate K_c for the reaction.
$$NO(g) + CO_2(g) \;\rightleftharpoons\; NO_2(g) + CO(g)$$

a) 0.033   b) 0.05   c) 0.25   d) 1.1   e) 0.33

41. Consider the reaction of carbon disulfide and chlorine according to the following equation:
$$CS_2(g) + 3\,Cl_2(g) \;\rightleftharpoons\; S_2Cl_2(g) + CCl_4(g)$$

When 0.80 mol of CS_2 and 2.4 mol of Cl_2 are placed in a 1.00 liter container and allowed to reach equilibrium, the mixture contains 0.60 mol CCl_4. What is the equilibrium concentration of Cl_2?

a) 0.60 mol/L   b) 1.2 mol/L   c) 1.6 mol/L   d) 1.8 mol/L   e) 2.1 mol/L

42. A mixture of 2.0 moles of N_2 and 2.0 moles of O_2 is placed in a 1.00 liter container at 448°C and allowed to reach equilibrium according to the following equation. The equilibrium constant is 100. Calculate the concentrations of all species present at equilibrium.

$$N_2(g) + O_2(g) \rightleftharpoons 2\,NO(g) \qquad K_c = 100.$$

a) [N_2] = 0.15   [O_2] = 0.15   [NO] = 1.5
b) [N_2] = 1.67   [O_2] = 1.67   [NO] = 0.34
c) [N_2] = 0.20   [O_2] = 0.20   [NO] = 2.5
d) [N_2] = 0.34   [O_2] = 0.34   [NO] = 3.3
e) [N_2] = 0.34   [O_2] = 0.34   [NO] = 1.7

43. A mixture of 0.40 moles of H_2 and 0.40 moles of I_2 is placed in a 1.00 liter container at 650°C and allowed to reach equilibrium according to the following equation. The equilibrium constant is 64. Calculate the concentrations of all species present at equilibrium.

$$H_2(g) + I_2(g) \rightleftharpoons 2\,HI\,(g) \qquad K = 64$$

a) [H_2] = 0.13   [I_2] = 0.13   [HI] = 0.53
b) [H_2] = 0.040   [I_2] = 0.040   [HI] = 0.32
c) [H_2] = 0.080   [I_2] = 0.080   [HI] = 0.64
d) [H_2] = 0.32   [I_2] = 0.32   [HI] = 0.24
e) [H_2] = 0.015   [I_2] = 0.015   [HI] = 0.12

44. A mixture of 0.50 moles of A and 0.50 moles of B is placed in a 1.00 liter container at 1100°C and allowed to reach equilibrium according to the following equation. The equilibrium constant is 0.16. Calculate the concentrations of all species present at equilibrium.

$$A(g) + B(g) \rightleftharpoons 2\,C(g)$$

a) [A] = 0.42   [B] = 0.42   [C] = 0.083
b) [A] = 0.38   [B] = 0.38   [C] = 0.24
c) [A] = 0.30   [B] = 0.30   [C] = 0.12
d) [A] = 0.15   [B] = 0.15   [C] = 0.060
e) [A] = 0.42   [B] = 0.42   [C] = 0.17

45. A 2.00 liter flask is filled with 1.5 mole SO_3, 2.5 moles SO_2, and 0.5 mole O_2, and allowed to reach equilibrium according to the following equation. At this temperature, $K_c = 4.00$. Predict the effect on the concentration of O_2 as equilibrium is being achieved by using Q, the reaction quotient?

$$2\ SO_{3(g)}\ \rightleftharpoons\ 2\ SO_{2(g)}\ +\ O_{2(g)}\qquad K = 4.00$$

   a) $[O_2]$ will increase because $Q < K$
   b) $[O_2]$ will increase because $Q > K$
   c) $[O_2]$ will decrease because $Q < K$
   d) $[O_2]$ will decrease because $Q > K$
   e) $[O_2]$ will remain the same because $Q=K$

46. Consider the reaction $2\ A(g)\ \rightleftharpoons\ B(g)$ where $K_c = 0.5$ at the temperature of the reaction. If 2.0 moles of A and 2.0 moles of B are introduced into a 1.00 liter flask, what change in concentrations (if any) would occur in time?
   a) [A] increases and [B] increases
   b) [A] increases and [B] decreases
   c) [A] decreases and [B] increases
   d) [A] decreases and [B] decreases
   e) [A] and [B] remain the same

47. Consider the reaction $A(g)\ \rightleftharpoons\ 2\ B(g)$ where $K_c = 1.5$ at the temperature of the reaction. If 3.0 moles of A and 3.0 moles of B are introduced into a 1.00 liter flask, what change in concentrations (if any) would occur in time?
   a) [A] increases and [B] increases
   b) [A] increases and [B] decreases
   c) [A] decreases and [B] increases
   d) [A] decreases and [B] decreases
   e) [A] and [B] remain the same

48. At a specific temperature, the equilibrium constant for the following reaction is given.
    $$2 SO_2(g) + O_2(g) \rightleftharpoons 2 SO_3(g) \qquad K_c = 1.2$$
    If 1.5 mol SO_2, 4.0 mol O_2 and 2.0 mol SO_3 are introduced into a 1.00 liter flask, what changes in concentration (if any) will be observed as the system reaches equilibrium?
    a) $[SO_2]$ increases; $[O_2]$ increases; $[SO_3]$ decreases
    b) $[SO_2]$ increases; $[O_2]$ decreases; $[SO_3]$ decreases
    c) $[SO_2]$ decreases; $[O_2]$ decreases; $[SO_3]$ increases
    d) $[SO_2]$ decreases; $[O_2]$ increases; $[SO_3]$ increases
    e) all concentrations remain the same

49. At a specific temperature, the equilibrium constant for the following reaction is given.
    $$2 NO_2(g) + O_2(g) \rightleftharpoons 2 NO_3(g) \qquad K_c = 0.25$$
    If 1.5 mol NO_2, 3.0 mol O_2 and 2.0 mol NO_3 are introduced into a 1.00 liter flask, what changes in concentration (if any) will be observed as the system reaches equilibrium?
    a) $[NO_2]$ increases; $[O_2]$ increases; $[NO_3]$ decreases
    b) $[NO_2]$ increases; $[O_2]$ decreases; $[NO_3]$ decreases
    c) $[NO_2]$ decreases; $[O_2]$ decreases; $[NO_3]$ increases
    d) $[NO_2]$ decreases; $[O_2]$ increases; $[NO_3]$ increases
    e) all concentrations remain the same

50. Exactly 0.50 mole of sulfur trioxide, 0.10 mole of sulfur dioxide, 0.20 mole of nitrogen monoxide and 0.30 mole nitrogen dioxide are sealed in a 1.0-L flask at 1500 °C. The equilibrium constant K_c is 0.24 for the following reaction.
    $$SO_3(g) + NO(g) \rightleftharpoons SO_2(g) + NO_2(g) \qquad K_c = 0.24$$
    When equilibrium is achieved, what changes in concentrations of SO_3 and NO will be observed?
    a) $[SO_3]$ increases; $[NO]$ increases
    b) $[SO_3]$ increases; $[NO]$ decreases
    c) $[SO_3]$ decreases; $[NO]$ decreases
    d) $[SO_3]$ decreases; $[NO]$ increases
    e) all concentrations remain the same

51. Consider the reaction of iodine and chlorine for which the enthalpy of reaction is –27 kJ.

$$I_2(aq) \; + \; Cl_2(aq) \; \rightleftharpoons \; 2 \; ICl(g) \qquad\qquad \Delta H = -27 \; kJ$$

At $25^{\circ}C$ $K_p = 1.6 \times 105$. If the temperature is increased to $100^{\circ}C$, what changes (if any) will be observed?

a) K_c will increase

b) no change because $K_c = K_p$

c) [ICl] will increase

d) [I_2] will increase

e) the partial pressure of ICl will increase

52. In which of the following reactions does a decrease in the volume of the container increase the concentration of the products? Assume constant temperature.

a) $\quad SO_2Cl_2(g) \; \rightleftharpoons \; SO_2(g) \; + \; Cl_2(g)$

b) $\quad C(s) \; + H_2O(g) \rightleftharpoons \; CO(g) \; + \; H_2(g)$

c) $\quad 2 \; SO_3 \; (g) \rightleftharpoons \; 2 \; SO_2(g) \; + \; O_2(g)$

d) $\quad I_2(g) \; + \; Cl_2(g) \; \rightleftharpoons \; 2 \; ICl \; (g)$

e) $\quad 2 \; NO \; + \; Br_2(g) \rightleftharpoons \; 2 \; NOBr(g)$

53. Ammonia is produced commercially by the Haber process in which nitrogen and hydrogen react by the reaction:

$$N_2(g) \; + \; 3 \; H_2(g) \; \rightleftharpoons \; 2 \; NH_3(g) \qquad\qquad \Delta H = -92.2 \; kJ$$

Which of the following changes, at equilibrium, will NOT result in a shift of the reaction to produce more NH_3?

a) removal of ammonia

b) addition of a catalyst

c) decreasing the size of the container

d) removal of hydrogen

e) increasing the temperature from $200^{\circ}C$ to $400^{\circ}C$

54. The reaction $A \rightleftharpoons 2 \; B$ is performed at two temperatures and the equilibrium constant determined. At temperature T_1, the $K = 800$. At temperature T_2 the $K = 20$. Which statement is true?

a) $T_1 > T_2$ and the reaction is endothermic.

b) $T_1 < T_2$ and the reaction is endothermic.

c) $T_1 > T_2$ and the reaction is exothermic.

d) T_2 is $40^{\circ}C$ less than T_1.

e) T_2 is 400 times greater than T_1.

55. A flask contains the following system at equilibrium:
$$PbCl_2(s) \rightleftharpoons Pb^{2+}(aq) + 2\,Cl^-(aq)$$

If solid NaCl is added to the system, what change (if any) will be observed?

a) more $PbCl_2$ will dissolve

b) more $PbCl_2$ will precipitate

c) more Pb^{2+} will be in solution

d) fewer Cl^- will be in solution

e) no change will be observed

56. A flask contains the following system at equilibrium:
$$Mg(OH)_2(s) \rightleftharpoons Mg^{2+}(aq) + 2\,OH^-(aq)$$

Which of the following reagents could be added to increase the solubility of $Mg(OH)_2$?

a) NaCl   b) NaOH   c) HCl   d) H_2O e) $MgCl_2$

57. Calcium carbonate decomposes when heated according to the following reaction:
$$CaCO_3(s) \rightleftharpoons CaO(s) + CO_2(g)$$

The mass of the $CaCO_3$ could be increased by

a) adding more CO_2

b) decreasing the volume of the container

c) removing some CaO

d) increasing the temperature

e) adding some CaO

58. In calculating concentrations at equilibrium, frequently one may assume that the molar concentration of a reagent changes very little ("x is small") and the initial concentration, $[A]_o$, is essentially equal to the equilibrium concentration. In which of the following cases is the assumption valid for the system
$$A \rightleftharpoons 2\,B \qquad\qquad K_c = 1.4 \times 10^{-3}$$

a) $[A]_o = 1.5 \times 10^{-3}$ b) $[A]_o = 4.7 \times 10^{-6}$ c) $[A]_o = 2.5 \times 10^{-1}$

d) $[A]_o = 5.1 \times 10^{-4}$ e) $[A]_o = 1.4 \times 10^{-4}$

59. In calculating concentrations at equilibrium, frequently one may assume that the molar concentration of a reagent changes very little ("x is small") and the initial concentration, $[A]_0$, is essentially equal to the equilibrium concentration. In which of the following cases is the assumption NOT valid for the system

$$A \rightleftharpoons 2B \qquad K_c = 7.1 \times 10^{-4}$$

a) $[A]_0 = 2.1$

b) $[A]_0 = 4.7 \times 10^1$

c) $[A]_0 = 3.6 \times 10^{-4}$

c) $[A]_0 = 8.1 \times 10^{-1}$

d) $[A]_0 = 1.4 \times 10^{-2}$

60. Which of the following statements concerning equilibrium is true?
   a) Catalysts are an effective means of changing the position of an equilibrium.
   b) The concentration of the products equals the concentration of reactions for a reaction a equilibrium.
   c) The equilibrium constant may be expressed in pressure terms or concentration terms for any reaction.
   d) When two opposing processes are proceeding at the same rate, the system is at equilibrium.
   e) A system at equilibrium cannot be disturbed.

Chapter 16 : Answers to Multiple Choice

11. c	21. b	31. b
12. c	22. c	32. c
13. d	23. a	33. d
14. c	24. a	34. a
15. d	25. b	35. b
16. d	26. e	36. e
17. c	27. b	37. d
18. c	28. b	38. e
19. c	29. d	39. a
20. a	30. c	40. c

41. a	51. d
42. d	52. e
43. c	53. d
44. e	54. a
45. a	55. b
46. e	56. c
47. b	57. b
48. c	58. c
49. a	59. c
50. c	60. d

Chapter 17
The Chemistry of Acids and Bases

Author's Note: Reference data is presented in tabular form such as a chemist actually uses. A student needs to be able to select the appropriate literature data for use in calculations as needed. Some users may prefer to present the constants within the problem.

Section A: Free Response

1. The pH of four solutions of known concentration was determined. Based on the results obtained, which sample is HCl? Explain your choice.

Sample	Concentration	pH
A	1.00×10^{-3} M	3.00
B	3.44 M	2.00
C	5.00×10^{-2} M	4.00
D	1.00×10^{-2} M	12.00

2. A chemist prepared a solution of HCl which was 1.00×10^{-9} M. What is the pH of the solution? Discuss the unusual characteristics of this system.

3. Methyl amine, CH_3NH_2 , is a weak base with $K_b = 5.0 \times 10^{-4}$ at 25°C. Write the equation for the hydrolysis of methyl amine in water and calculate the pH of a 1.12 M solution of the compound.

4. Predict whether the equilibrium lies predominately to the left or to the right in the reaction below. Explain your choice.

    a) $HCO_3^-(aq)$ + $H_2O(l)$ $\rightleftharpoons$ $CO_3^{2-}(aq)$ + $H_3O^+(aq)$

    b) $C_5H_5N(aq)$ + $H_3O^+(aq)$ $\rightleftharpoons$ $C_5H_5NH^+(aq)$ + $H_2O(l)$

Reference

	Acid	K_a	Conj. Base	K_b
Hydrofluoric Acid	HF	7.2×10^{-4}	F^-	1.4×10^{-11}
Nitrous Acid	HNO_2	4.5×10^{-4}	NO_2^-	2.2×10^{-11}
Acetic Acid	CH_3CO_2H	1.8×10^{-5}	$CH_3CO_2^-$	5.6×10^{-10}
Benzoic Acid	$C_6H_5CO_2H$	6.3×10^{-5}	$C_6H_5CO_2^-$	1.6×10^{-10}
Ammonium ion	NH_4^+	5.6×10^{-10}	NH_3	1.8×10^{-5}
Hydrocyanic acid	HCN	4.0×10^{-10}	CN^-	2.5×10^{-5}

5. Explain why a 0.1 M solution of NaCl is neutral and a 0.1 M solution of NaF is basic.

6. Calcium hydroxide is a slightly soluble compound. Does the pH of a saturated calcium hydroxide solution increase, decrease, or remain the same when calcium chloride is also present? Explain.

7. Consider a 0.877 M solution of potassium benzoate, $KC_6H_5CO_2$. What is the spectator ion in the solution? Write the equation for any hydrolysis reaction that might occur in this solution and predict if the solution should be acidic, basic or neutral. Calculate the pH of the solution.

8. Write the equations for the reaction of sulfuric acid in water. Discuss the degree to which each occurs.

9. Predict the relative strengths of chloroacetic acid and trichloroacetic acid as compared to acetic acid. Explain the trend in acidities.

CH_3CO_2H $ClCH_2CO_2H$ Cl_3CCO_2H
acetic acid   chloroacetic acid   trichloroacetic acid

10. Write equations to illustrate the amphoteric nature of the compound $Zn(OH)_2$ based on the following experiments. Discuss the nature of each reaction.
   Reaction #1: Sodium hydroxide is added to a solution of zinc chloride. A white precipitate forms. The precipitate is used for Reaction 2 and Reaction 3.
   Reaction #2: Acid is added to the white precipitate. It dissolves.
   Reaction #3: Base is added to the white precipitate. It dissolves.

Key Concepts for Free Response:

1. Sample A is HCl. Since HCl is a strong acid the $[H_3O^+]$ would be 0.001 and a pH of 3. All the other samples are only partially ionized and cannot be HCl.

2. The pH of the solution is 7.00. Although the chemist added just a few drops of HCl to a large amount of water so the $[H_3O^+]$ was 1.00×10^{-9}, water has a leveling effect on the solution. The strongest acid that can exist in water is H3O+ so 1×10^{-7} is the limit.

3. $CH_3NH_2 + H_2O \rightleftharpoons CH_3NH_3^+ + OH^-$ For a 1.12 M solution, pH = 10.75

4. a) The equilibrium lies predominately to the left because HCO_3^- is a weaker acid than H_3O^+ and H_2O is a weaker base than CO_3^{2-}. All proton transfer reaction go toward the weaker acid/base pair.

   b) The equilibrium lies predominately to the right because $C_5H_5NH^+$ is a weaker acid than H_3O^+ and H_2O is a weaker base than C_5H_5N.

5. Sodium chloride dissociates into Na^+ and Cl^- neither of which react with water. Thus the solution has a neutral pH. The F^-, however, reacts with water according to

$$F^-(aq) + H_2O(l) \rightleftharpoons HF(aq) + OH^-(aq)$$

6. Calcium chloride provides additional Ca^{2+} which shifts the equilibrium to the left.

$$Ca(OH)_2(s) \rightleftharpoons Ca^{2+}(aq) + 2\,OH^-(aq)$$

   Therefore the $[OH^-]$ is decreased and the initial basic pH is nearer 7; the pH is decreased.

7. The spectator ion is K^+ which does not enter into the reaction. A basic solution is predicted from the hydrolysis reaction is

$$C_6H_5CO_2^-(aq) + H_2O(l) \rightleftharpoons C_6H_5CO_2H(aq) + OH^-(aq)$$

   To calculate the pH, use Kb. pH = 9.07

8. The reactions of sulfuric acid in water

$$H_2SO_4 \quad + \quad H_2O(l) \longrightarrow \quad HSO_4^-(aq) + H_3O^+(aq) \quad K_{a1} = \text{very large}$$

   In water this reaction is 100% to the ionized form.

$$HSO_4^-(aq) \quad + \quad H_2O(l) \rightleftharpoons \quad SO_4^{2-}(aq) + H_3O^+(aq) \qquad K_{a2} = \text{smaller}$$

   this reaction occurs to a lesser extent

9. Chloroacetic acid would be more acidic than acetic acid because the electronegative chlorine atom draws electrons toward it. This makes it easier for the H+ to be transferred. The effect is even greater for the trichloroacetic acid as three chlorine atoms are withdrawing electrons. Trichloroacetic acid would be much stronger than the other two acids.

10. Reaction #1: Formation of solid zinc hydroxide

$$ZnCl_2(aq) + 2\,NaOH(aq) \longrightarrow Zn(OH)_2(s) + NaCl(aq)$$

   Reaction #2: Zinc hydroxide is reacting as a base

$$Zn(OH)_2(s) + 2\,H_3O^+(aq) \rightleftharpoons Zn^{2+}(aq) + 4\,H_2O(l)$$

   base   acid   soluble compound

   Reaction #3: Zinc hydroxide is reacting as an acid

$$Zn(OH)_2(s) + 2\,OH^-(aq) \rightleftharpoons [Zn(OH)_2]^{2-}(aq)$$

   acid   base   a soluble complex ion of a transition metal

Section B: Multiple Choice

Author's Note: Reference data is presented in tabular form such as a chemist actually uses. A student needs to be able to select the appropriate literature data for use in calculations as needed. Some users may prefer to present the constants within the problem.

Reference

Acid		K_a	Conj. Base	K_b
Hydrofluoric Acid	HF	7.2×10^{-4}	F^-	1.4×10^{-11}
Nitrous Acid	HNO_2	4.5×10^{-4}	NO_2^-	2.2×10^{-11}
Acetic Acid	CH_3CO_2H	1.8×10^{-5}	$CH_3CO_2^-$	5.6×10^{-10}
Ammonium ion	NH_4^+	5.6×10^{-10}	NH_3	1.8×10^{-5}
Hydrocyanic acid	HCN	4.0×10^{-10}	CN^-	2.5×10^{-5}

11. In the equation, $HF(aq) + H_2O(l) \rightleftharpoons H_3O^+(aq) + F^-(aq)$

    a) HF is an acid and H_3O^+ is its conjugate base.

    b) H_2O is an acid and H_3O^+ is its conjugate base.

    c) HF is an acid and F^- is its conjugate base.

    d) H_2O is an acid and H_3O^+ is its conjugate base.

    e) HF is an acid and H_2O is its conjugate base.

12. In the equation, $NH_4^+(aq) + H_2O(l) \rightleftharpoons NH_3(aq) + H_3O^+(aq)$

    a) NH_4^+ is an acid and NH_3 is its conjugate base.

    b) H_2O is an acid and H_3O^+ is its conjugate base.

    c) NH_4^+ is an acid and H_3O^+ is its conjugate base.

    d) H_2O is an acid and NH_4^+ is its conjugate base.

    e) NH_3 is an acid and NH_4^+ is its conjugate base.

13. In the equation, $HCO_3^-(aq) + H_2O(l) \rightleftharpoons CO_3^{2-}(aq) + H_3O^+(aq)$

    a) HCO_3^- is an base and H_3O^+ is its conjugate acid.

    b) HCO_3^- is an base and CO_3^{2-} is its conjugate acid.

    c) H_2O is an base and H_3O^+ is its conjugate acid.

    d) H_2O is an base and CO_3^{2-} is its conjugate acid.

    e) H_2O is an base and CO_3^{2-} is its conjugate acid.

Reference

	Acid	K_a	Conj. Base	K_b
Hydrofluoric Acid	HF	7.2×10^{-4}	F^-	1.4×10^{-11}
Nitrous Acid	HNO_2	4.5×10^{-4}	NO_2^-	2.2×10^{-11}
Acetic Acid	CH_3CO_2H	1.8×10^{-5}	$CH_3CO_2^-$	5.6×10^{-10}
Ammonium ion	NH_4^+	5.6×10^{-10}	NH_3	1.8×10^{-5}
Hydrocyanic acid	HCN	4.0×10^{-10}	CN^-	2.5×10^{-5}

14. In the equation, $NO_2^-(aq) + H_2O(l) \rightleftharpoons HNO_2(aq) + OH^-(aq)$

    a) H_2O is a base and OH^- is its conjugate acid.

    b) H_2O is a base and HNO_2 is its conjugate acid.

    c) NO_2^- is a base and HNO_2 is its conjugate acid.

    d) NO_2^- is a base and OH^- is its conjugate acid.

    e) NO_2^- is a base and H_2O is its conjugate acid.

15. The conjugate acid of OH^- is

    a) O^{2-} b) H_2O c) H_3O^+ d) H^+ e) O_2^-

16. The conjugate base of HCO_3^- is

    a) CO_3^{2-} b) H_2CO_3 c) H_3O^+ d) OH^- e) H_2O

17. The conjugate acid of SO_3^{2-} is

    a) SO_4^{2-} b) HSO_3^- c) SO_3 d) H^+ e) H_3O^+

18. The species which is the conjugate acid to the HCO_3^- ion is

    a) CO_2 b) CO_3^{2-} c) H_3O^+ d) H_2O e) H_2CO_3

19. Which of the following solutions will have a pH of 1?

    a) 1.0 M CH_3CO_2H b) 0.1 M CH_3CO_2H c) 0.1 M HF

    d) 0.1 M HNO_3 e) 0.1 M NH_3

20. Which of the following is a strong acid?

    a) CH_3CO_2H b) HNO_3 c) HNO_2 d) HF   e) $B(OH)_3$

21. Which of the following is a weak base?

    a) KOH   b) $B(OH)_3$ c) NH_3 d) CH_3CO_2H e) $Mg(OH)_2$

Reference

	Acid	K_a	Conj. Base	K_b
Hydrofluoric Acid	HF	7.2×10^{-4}	F^-	1.4×10^{-11}
Nitrous Acid	HNO_2	4.5×10^{-4}	NO_2^-	2.2×10^{-11}
Acetic Acid	CH_3CO_2H	1.8×10^{-5}	$CH_3CO_2^-$	5.6×10^{-10}
Ammonium ion	NH_4^+	5.6×10^{-10}	NH_3	1.8×10^{-5}
Hydrocyanic acid	HCN	4.0×10^{-10}	CN^-	2.5×10^{-5}

22. At 50^oC the water ionization constant, K_w, is 5.48×10^{-14}. What is the $[H_3O^+]$ in neutral water at 50^oC?

    a) 5.48×10^{-7} b) 2.34×10^{-7} c) 1.00×10^{-7}
    d) 5.48×10^{-14} e) 4.27×10^{-13}

23. At 15^oC the water ionization constant, K_w, is 0.45×10^{-14}. What is the $[H_3O^+]$ in neutral water at 15^oC?

    a) 6.71×10^{-8} b) 1.50×10^{-7} c) 1.00×10^{-7}
    d) 4.5×10^{-15} e) 5.50×10^{-13}

24. Which of the following solutions will have a pH of 13?

    a) 1.0 M NH_3 b) 0.1 M HNO_3 c) 0.1 M CH_3CO_2H
    d) 0.1 M HF   e) 0.1 M KOH

25. What is the pH of a 4.20×10^{-4} M HBr solution at 25^oC?

    a) 2.80 b) 3.38 c) 3.80 d) 4.20 e) 4.62

26. What is the pH of a 0.0813 M HNO_3 solution at 25^oC?

    a) 0.813 b) 0.910 c) 1.090 d) 1.813 e) 1.870

Reference

Acid		K_a	Conj. Base	K_b
Hydrofluoric Acid	HF	7.2×10^{-4}	F^-	1.4×10^{-11}
Nitrous Acid	HNO_2	4.5×10^{-4}	NO_2^-	2.2×10^{-11}
Acetic Acid	CH_3CO_2H	1.8×10^{-5}	$CH_3CO_2^-$	5.6×10^{-10}
Ammonium ion	NH_4^+	5.6×10^{-10}	NH_3	1.8×10^{-5}
Hydrocyanic acid	HCN	4.0×10^{-10}	CN^-	2.5×10^{-5}

27. After consulting the reference chart, select the stronger acid in each set.

Set #1: $HC_2H_3O_2$ or HNO_2

Set #2: NH_4^+ or H_2O

Set #3: HCO_3^- or H_2CO_3

a) $HC_2H_3O_2$, H_2CO_3, H_2O b) HNO_2, HCO_3^-, H_2O

c) HNO_2, H_2CO_3, NH_4^+ d) $HC_2H_3O_2$, H_2CO_3, NH_4^+

e) $HC_2H_3O_2$, HCO_3^-, H_2O

28. After consulting the reference chart, what is the stronger base in each set?

Set #1: $C_2H_3O_2^-$ or NO_2^-

Set #2: NH_3 or OH^-

Set #3: HCO_3^- or CO_3^{2-}

a) NO_2^-, CO_3^{2-}, NH_3 b) NO_2^-, HCO_3^-, OH^-

c) NO_2^-, CO_3^{2-}, NH_3 d) $C_2H_3O_2^-$, HCO_3^-, NH_3

e) $C_2H_3O_2^-$, CO_3^{2-}, OH^-

29. What is the pH of a 3.18 M CH_3CO_2H solution at 25°C? $K_a = 1.8 \times 10^{-5}$

a) 2.12   b) 2.75   c) 1.40   d) 4.24   e) 4.74

30. What is the pH of a 1.86 M $CH_3CH_2CO_2H$ solution at 25°C? $K_a = 1.3 \times 10^{-5}$

a) 4.92   b) 4.88   c) 2.42   d) 2.31   e) 2.08

31. What is the pH of a 3.51 M HCN solution at 25°C? $K_a = 4.0 \times 10^{-10}$

a) 1.00   b) 1.48   c) 3.75   d) 4.43   e) 7.00

32. What is the pH of a 4.24 M $C_6H_5CO_2H$ solution at 25°C? $K_a = 6.3 \times 10^{-5}$

a) 2.08   b) 2.31   c) 2.42   d) 4.88   e) 4.92

Reference

	Acid	K_a	Conj. Base	K_b
Hydrofluoric Acid	HF	7.2×10^{-4}	F^-	1.4×10^{-11}
Nitrous Acid	HNO_2	4.5×10^{-4}	NO_2^-	2.2×10^{-11}
Acetic Acid	CH_3CO_2H	1.8×10^{-5}	$CH_3CO_2^-$	5.6×10^{-10}
Ammonium ion	NH_4^+	5.6×10^{-10}	NH_3	1.8×10^{-5}
Hydrocyanic acid	HCN	4.0×10^{-10}	CN^-	2.5×10^{-5}

33. What is the pH of a 0.0536 M NaOH solution at 25°C?

    a) 1.14   b) 1.27   c) 8.64   d) 12.73   e) 13.95

34. What is the pH of a 0.0144 M $Ca(OH)_2$ solution at 25°C?

    a) 1.54   b) 1.84   c) 10.84   d) 12.16   e) 12.45

35. What is the pH of a 0.0443 M ammonia (NH_3) solution at 25°C? $K_b = 1.8 \times 10^{-5}$

    a) 3.05   b) 6.10   c) 9.25   d) 10.95   e) 12.64

36. What is the pH of a 2.54 M $NH_2(CH_3)$ solution at 25°C? $K_b = 5.0 \times 10^{-4}$

    a) 1.45   b) 3.35   c) 10.70   d) 11.10   e) 12.52

37. The pH of a solution at 25°C is 4.14. The $[H_3O^+]$ is

    a) 1.4×10^{-4} M   b) 7.8×10^{-3} M   c) 1.0×10^{-3} M

    d) 2.4×10^{-8} M   e) 1.6×10^{-10} M

38. The pH of a solution at 25°C is 9.14. The $[OH^-]$ is

    a) 1.4×10^{-5} M   b) 3.7×10^{-7} M   c) 8.5×10^{-8} M

    d) 1.4×10^{-9} M   e) 7.2×10^{-10} M

39. The pH of a solution at 25°C is 11.86. The $[OH^-]$ is

    a) 6.6×10^{-2} M   b) 7.2×10^{-3} M   c) 2.1×10^{-10} M

    d) 1.4×10^{-12} M   e) 1.5×10^{-13} M

40. The pH of a solution at 25°C is 4.13. The $[H_3O^+]$ is

    a) 7.4×10^{-5} M   b) 7.4×10^{-3} M   c) 8.9×10^{-4} M

    d) 1.4×10^{-10} M   e) 1.2×10^{-12} M

Reference

	Acid	K_a	Conj. Base	K_b
Hydrofluoric Acid	HF	7.2×10^{-4}	F^-	1.4×10^{-11}
Nitrous Acid	HNO_2	4.5×10^{-4}	NO_2^-	2.2×10^{-11}
Acetic Acid	CH_3CO_2H	1.8×10^{-5}	$CH_3CO_2^-$	5.6×10^{-10}
Ammonium ion	NH_4^+	5.6×10^{-10}	NH_3	1.8×10^{-5}
Hydrocyanic acid	HCN	4.0×10^{-10}	CN^-	2.5×10^{-5}

41. The pH of a 2.28 M solution of a weak acid is 5.21 at 25°C. What is K_a for the weak acid?

a) 2.1×10^{-5} b) 6.2×10^{-6} c) 8.8×10^{-9}

d) 4.8×10^{-10} e) 1.7×10^{-11}

42. The pH of a 4.52 M solution of a weak acid is 3.90 at 25°C. What is K_a for the weak acid?

a) 1.3×10^{-4} b) 1.8×10^{-5} c) 2.9×10^{-6}

d) 3.5×10^{-9} e) 1.5×10^{-11}

43. The pH of a 2.10 M solution of a weak base is 12.57 at 25°C. What is K_b for the weak base?

a) 3.7×10^{-2} b) 6.6×10^{-4} c) 3.9×10^{-6}

d) 1.5×10^{-11} e) 2.7×10^{-13}

44. The pH of a 4.82 M solution of a weak base is 10.35 at 25°C. What is K_b for the weak base?

a) 2.2×10^{-4} b) 9.6×10^{-7} c) 1.0×10^{-8}

d) 5.0×10^{-10} e) 4.4×10^{-11}

45. What is the % ionization of a 3.14 M CH_3CO_2H solution at 25°C? For CH_3CO_2H, $K_a = 1.8 \times 10^{-8}$.

a) 0.24%   b) 0.57%   c) 1.8%   d) 3.2%   e) 7.5%

46. What is the % ionization of a 2.50 M NH_3 solution at 25°C? For NH_3, $K_b = 1.8 \times 10^{-8}$.

a) 0.0085%   b) 1.5%   c) 2.1%   d) 3.9%   e) 7.3%

Reference

	Acid	K_a	Conj. Base	K_b
Hydrofluoric Acid	HF	7.2×10^{-4}	F^-	1.4×10^{-11}
Nitrous Acid	HNO_2	4.5×10^{-4}	NO_2^-	2.2×10^{-11}
Acetic Acid	CH_3CO_2H	1.8×10^{-5}	$CH_3CO_2^-$	5.6×10^{-10}
Ammonium ion	NH_4^+	5.6×10^{-10}	NH_3	1.8×10^{-5}
Hydrocyanic acid	HCN	4.0×10^{-10}	CN^-	2.5×10^{-5}

47. Three weak acids have the formulas and K_a values listed.

           Formic acid CHO_2H 1.8×10^{-4}

           Cyanic Acid   HOCN 3.5×10^{-4}

           Chloroacetic Acid $C_2H_2ClO_2H$ 1.4×10^{-3}

     Which of the following is the strongest base?

     a) $C_2H_2ClO_2^-$ b) OCN^- c) HCO_2^- d) H_2O e) H_3O^+

48. Of the salts, NaF, NaCl, NaBr, $NaNO_2$, and $NaNO_3$, how many would form a neutral solution?

     a) four   b) three   c) two   d) one   e) zero (none form neutral solutions)

49. Of the salts, KCH_3CO_2, KF, KCN, and KNO_2, how many would form a basic aqueous solution?

     a) four   b) three   c) two   d) one   e) zero (none form basic solutions)

50. Of the salts, KCH_3CO_2, NH_4Cl, KBr, and NH_4NO_3, how many would form an acidic aqueous solution?

     a) four   b) three   c) two   d) one   e) zero (none form basic solutions)

51. Of the salts, $NaCH_3CO_2$, NaCN, NaF, and NH_4NO_3, how many would form a neutral solution?

     a) four   b) three   c) two   d) one   e) zero (none form neutral solutions)

52. Of the following salts, which one forms a 0.1 M solution with the highest pH?

     a) NH_4Cl b) KCH_3CO_2 c) KBr   d) $NaNO_2$ e) NaCl

53. Of the following salts, which one forms a 0.1 M solution with the lowest pH?

     a) NH_4Cl b) KCH_3CO_2 c) KBr   d) $NaNO_2$ e) NaCl

Reference

	Acid	K_a	Conj. Base	K_b
Hydrofluoric Acid	HF	7.2×10^{-4}	F^-	1.4×10^{-11}
Nitrous Acid	HNO_2	4.5×10^{-4}	NO_2^-	2.2×10^{-11}
Acetic Acid	CH_3CO_2H	1.8×10^{-5}	$CH_3CO_2^-$	5.6×10^{-10}
Ammonium ion	NH_4^+	5.6×10^{-10}	NH_3	1.8×10^{-5}
Hydrocyanic acid	HCN	4.0×10^{-10}	CN^-	2.5×10^{-5}

54. At 25°C what is the pH of a 2.50 M solution of sodium benzoate, $NaC_6H_5CO_2$?

   a) 1.90   b) 4.70   c) 9.30   d) 9.80   e) 12.10

55. At 25°C what is the pH of a 1.75 M solution of sodium cyanide NaCN?

   a) 11.82   b) 10.04   c) 3.44   d) 1.18   e) 0.80

56. At 25°C what is the pH of a 3.25 M solution of ammonium chloride, NH_4Cl?

   a) 2.37   b) 4.37   c) 4.62   d) 9.37   e) 9.63

57. At 25°C what is the pH of a 0.084 M solution of potassium nitrite, KNO_2?

   a) 4.13   b) 5.87   c) 6.25   d) 8.13   e) 9.87

58. Which of the following salts, when dissolved in water solution, will give the most acidic solution?

   a) $MgBr_2$ b) KNO_3 c) LiCl   d) Na_2SO_4 e) $FeBr_3$

59. Using the Lewis acid-base concept, predict which of the metal ions below will produce the most acidic aqueous solution.

   a) Na^+ b) Mg^{2+} c) Al^{3+} d) Co^{2+} e) Li^+

60. Which of the following compounds will form an acidic aqueous solution?

   a) CO_2 b) CaO   c) $Mg(OH)_2$ d) NaBr   e) $NaCH_3CO_2$

61. Which of the following compounds would form a 0.1 M solution with a pH of about 5?

   a) NH_3 b) NaCl   c) NH_4NO_3 d) HCl   e) NaBr

Reference

	Acid	K_a	Conj. Base	K_b
Hydrofluoric Acid	HF	7.2×10^{-4}	F^-	1.4×10^{-11}
Nitrous Acid	HNO_2	4.5×10^{-4}	NO_2^-	2.2×10^{-11}
Acetic Acid	CH_3CO_2H	1.8×10^{-5}	$CH_3CO_2^-$	5.6×10^{-10}
Ammonium ion	NH_4^+	5.6×10^{-10}	NH_3	1.8×10^{-5}
Hydrocyanic acid	HCN	4.0×10^{-10}	CN^-	2.5×10^{-5}

62. Water cannot function as which one of the following?

a) a conjugate base   b) a Brønsted acid   c) a Brønsted base

d) a Lewis acid   e) a Lewis base

63. Oxalic acid, $H_2C_2O_4$, is a weak diprotic acid. In a 0.1 M solution of oxalic acid, which species would have the lowest concentration?

a) $H_2C_2O_4$ b) $HC_2O_4^-$ c) H_3O^+ d) $C_2O_4^{2-}$ e) H_2O

64. Oxalic acid, $H_2C_2O_4$, is a diprotic acid with $K_1 = 5.6 \times 10^{-2}$ and $K_2 = 5.1 \times 10^{-5}$. What is the equilibrium constant for the following reaction?

$$H_2C_2O_4,(aq) + 2 H_2O(l) \rightleftharpoons 2 H_3O^+(aq) + C_2O_4^{2-}(aq)$$

a) 5.6×10^{-2} b) 3.1×10^{-3} c) 2.9×10^{-6}

d) 3.5×10^{-5} e) 1.8×10^{-13}

65. All of the following compounds are acids containing chlorine. Which compound is the strongest acid?

a) $HClO_4$ b) $HClO_3$ c) $HClO_2$ d) $HClO$ e) HCl

Chapter 17 : Answers to Multiple Choice

11. c	21. c	31. d
12. a	22. b	32. b
13. c	23. a	33. d
14. c	24. e	34. e
15. b	25. b	35. d
16. a	26. c	36. e
17. b	27. c	37. a
18. e	28. e	38. e
19. d	29. a	39. d
20. b	30. d	40. a

41. e	51. e	61. c
42. d	52. b	62. d
43. b	53. a	63. d
44. c	54. c	64. c
45. a	55. a	65. a
46. a	56. b	
47. c	57. d	
48. c	58. e	
49. a	59. c	
50. c	60. a	

Chapter 18
Reactions of Acids and Bases

Author's Note: Reference data is presented in tabular form such as a chemist actually uses. A student needs to be able to select the appropriate literature data for use in calculations as needed. Some users may prefer to present the constants within the problem.

Section A: Free Response

Reference

	Acid	K_a	Conj. Base	K_b
Hydrofluoric Acid	HF	7.2×10^{-4}	F^-	1.4×10^{-11}
Nitrous Acid	HNO_2	4.5×10^{-4}	NO_2^-	2.2×10^{-11}
Acetic Acid	CH_3CO_2H	1.8×10^{-5}	$CH_3CO_2^-$	5.6×10^{-10}
Ammonium ion	NH_4^+	5.6×10^{-10}	NH_3	1.8×10^{-5}
Hydrocyanic acid	HCN	4.0×10^{-10}	CN^-	2.5×10^{-5}

1. Sodium acetate and nitrous acid are mixed in water. Write a balanced equation for the acid-base reaction that could, in principle, occur. Does the reaction occur to any significant extent? Explain why or why not.

2. Write the chemical equation to represent an aqueous solution of hydrofluoric acid, HF. Use LeChatelier's principle to discuss the change in pH (if any) that would be observed when solid sodium fluoride is added to the above solution of hydrofluoric acid.

3. Exactly 50.0 mL of 0.30 M HCN is added to 50.0 mL of 0.30 M NaOH. Predict and then calculate the pH of the resulting solution?

4. What is the pH of a titration mixture when 15.0 mL of 0.250 M NaOH has been added to 25.0 mL of 0.400 M CH_3CO_2H? Predict the position on the titration curve for this mixture.

5. What is the pH of a titration mixture when 20.0 mL of 0.250 M NaOH has been added to 25.0 mL of 0.400 M CH_3CO_2H? Predict the position on the titration curve for this mixture.

6. What is the pH of a titration mixture when 40.0 mL of 0.250 M NaOH has been added to 25.0 mL of 0.400 M CH_3CO_2H? Predict the position on the titration curve for this mixture.

7. To prepare a buffer of pH 8—10, which reagent, NH_4Cl, CH_3CO_2H or NaOH, should be added to a solution of ammonia? Explain your choice.

	Acid	K_a	Conj. Base	K_b
Hydrofluoric Acid	HF	7.2×10^{-4}	F^-	1.4×10^{-11}
Nitrous Acid	HNO_2	4.5×10^{-4}	NO_2^-	2.2×10^{-11}
Acetic Acid	CH_3CO_2H	1.8×10^{-5}	$CH_3CO_2^-$	5.6×10^{-10}
Ammonium ion	NH_4^+	5.6×10^{-10}	NH_3	1.8×10^{-5}
Hydrocyanic acid	HCN	4.0×10^{-10}	CN^-	2.5×10^{-5}

8. The titration curve shows the reaction of 0.1 M solutions of two of the reagents, HCl, NaOH, CH_3CO_2H and NH_3. Select the two that were used and discuss the significance of the indicated points on the curve.

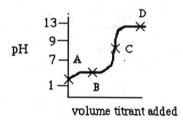

9. Sketch the titration curve that would represent a flask of 0.10 M NH_3 being treated with 0.10 M HCl from a buret. Label and discuss the significance of the initial pH, a point early in the titration, and the equivalence point.

10. Phenolphthalein is a useful indicator for acid/base reaction. A student placed a few drops of indicator in an acid solution and observed no color. The base was placed in a buret and added to the acid until a permanent pink color was observed. The volume recorded was used to calculate the equivalence point of the titration. Would phenolphthalein be as useful if the titration mixtures were reversed, that is the base and indicator were in the flask and the acid was in the buret? Explain.

Key Concepts for Free Response:

1. The acetate ion could react with the unionized nitrous acid according to the equation.

    $$CH_3CO_2^-(aq) + HNO_2(aq) \rightleftharpoons CH_3CO_2H(aq) + NO_2^- aq)$$

    The reaction would occur because CH_3CO_2H is a weaker acid than HNO_2 and NO_2^- is a weaker base than $CH_3CO_2^-$. The equilibrium favors the products.

2. The reaction is

    $$HF(aq) + H_2O(aq) \rightleftharpoons F^-(aq) + H_3O^+(aq)$$

    When solid NaF is added, the F- causes the equilibrium to shift to the left and the $[H_3O^+]$ decreases. This would correspond to an increase in the pH.

3. Equal molar quantities of acid and base are added, so the mixture is at the equivalence point. Thus it is merely an aqueous solution of NaCN which would be basic according to the equation

    $$CN^-(aq) + H_2O(aq) \rightleftharpoons HCN(aq) + OH^-(aq)$$

    The concentration of CN^- is 0.015 mol/0.100L = 0.15 M. The pH is calculated as 11.29.

4. The mixture is early in the titration because only 0.00375 mole of base were added forming 0.00375 moles of salt in a volume of 0.040 L. The pH is 4.52.

5. The mixture is at the half-neutralization point because 0.05 moles of base have been added. The pH is 4.74.

6. The mixture is at the equivalence point because 0.10 moles of base have been added. We predict is to be slight basic because it is merely a solution of the salt.
    0.010 mole $CH_3CO_2^-$ / 0.065 L = 0.154 M. The pH is 8.97.

7. NH_4Cl could be added to a solution of NH_3 to give a basic buffer. The equations are

    $$NH_3(aq) + H_2O(aq) \rightleftharpoons NH_4^+(aq) + OH^-(aq)$$

    $$NH_4^+(aq) + H_2O(aq) \rightleftharpoons NH_3(aq) + H_3O^+(aq)$$

    When base is added the NH_4^+ ion reacts, and when acid is added the NH_3 molecule reacts. This causes the pH to remain fairly constant within the range.

8. The titration is performed with CH_3CO_2H in the flask and NaOH in the buret. At point A the pH is not 1.0 so the acid is weak. At point B, the solution is half-neutralize and the pH is 4.74. The "flat" region of the curve shows the buffering action of the mixture of unreacted CH_3CO_2H and the newly formed $CH_3CO_2^-$. At point C the solution is neutralized and the pH is 8.25. At point D the excess NaOH shows a high pH nearing 13.

9. Initially the pH is less than 13 because the NH_3 is partially ionized. As HCl is added the pH decreases. At the half-neutralization point the pH is 9.26. At the equivalence point there is merely a solution of NH_4Cl which has a slightly acidic pH.

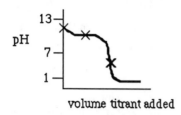

10. In a titration, it is best to observe a sharp endpoint where one drop of solution causes a noticeable difference. Phenolphthalein is a good indicator because the change is colorless in acid to pink in base. The first drop of excess base causes the appearance of a new color. If the titration were reversed, in theory the same reaction takes place and the color would disappear. In practice, though, seeing a color disappear is much more difficult to observe than seeing a color appear. Another indicator would be advised.

Section B: Multiple Choice
Author's Note: Reference data is presented in tabular form such as a chemist actually uses. A student needs to be able to select the appropriate literature data for use in calculations as needed. Some users may prefer to present the constants within the problem.

Reference

Acid		K_a	Conj. Base	K_b
Hydrofluoric Acid	HF	7.2×10^{-4}	F^-	1.4×10^{-11}
Nitrous Acid	HNO_2	4.5×10^{-4}	NO_2^-	2.2×10^{-11}
Acetic Acid	CH_3CO_2H	1.8×10^{-5}	$CH_3CO_2^-$	5.6×10^{-10}
Ammonium ion	NH_4^+	5.6×10^{-10}	NH_3	1.8×10^{-5}
Hydrocyanic acid	HCN	4.0×10^{-10}	CN^-	2.5×10^{-5}

11. What are the products of the following acid-base reaction?

$$NaOH(aq) + HF \longrightarrow$$

a) $NaF(aq)$ and $H_2O(l)$ b) NaH (aq) and $HOF(aq)$ c) H_2O only

d) $H_2O(l)$, $F_2(aq)$, and Na (s)   e) $NaF(aq) + OH^-(aq)$

12. What are the products of the following acid-base reaction?

$$NaOH(aq) + HNO_3(aq) \longrightarrow$$

a) $NaH(aq)$ and $H_2O(l)$ b) $NaNO_3$ (aq) and $H_2O(l)$ c) H_2O only

d) $H_2O(l)$, $N_2(g)$, and Na(s)   e) $NaNO_3(aq) + H_3O^+$

13. What are the products of the following acid-base reaction?

$$CH_3CO_2H + KOH(aq) \longrightarrow$$

a) CH_3CO_2H (aq) and $H_2O(l)$ b) $KH(aq)$ and CH_3CO_2H (aq)   c) H_2O only

d) $H_2O(l)$, $CO_2,(g)$, and K(s)   e) $KCH_3CO_2(aq)$ and $H_2O(l)$

	Acid	Ka	Conj. Base	Kb
Hydrofluoric Acid	HF	7.2×10^{-4}	F^-	1.4×10^{-11}
Nitrous Acid	HNO_2	4.5×10^{-4}	NO_2^-	2.2×10^{-11}
Acetic Acid	CH_3CO_2H	1.8×10^{-5}	$CH_3CO_2^-$	5.6×10^{-10}
Ammonium ion	NH_4^+	5.6×10^{-10}	NH_3	1.8×10^{-5}
Hydrocyanic acid	HCN	4.0×10^{-10}	CN^-	2.5×10^{-5}

14. Which of the following acid-base reactions will lie predominantly toward the products?

   Reaction #1: $HF(aq) + NH_3(aq) \rightleftharpoons NH_4^+(aq) + F^-(aq)$

   Reaction #2: $NH_3(aq) + H_2O(l) \rightleftharpoons NH_4^+(aq) + OH^-(aq)$

   Reaction #3: $HF(aq) + H_2O(l) \rightleftharpoons H_3O^+(aq) + F^-(aq)$

   a) #1 only   b) #2 only   c) #1 and #2

   d) #2 and #3   e) all three reactions

15. Which of the following acid-base reactions will lie predominantly toward the products?

   Reaction #1: $NH_3(aq) + H_2O(l) \rightleftharpoons NH_4^+(aq) + OH^-(aq)$

   Reaction #2: $CH_3CO_2H(aq) + H_2O(l) \rightleftharpoons H_3O^+(aq) + CH_3CO_2^-(aq)$

   Reaction #3: $CH_3CO_2H(aq) + NH_3(aq) \rightleftharpoons NH_4^+(aq) + CH_3CO_2^-(aq)$

   a) #1 only   b) #2 only   c) #3 only

   d) #1 and #2   e) #1 and #3

16. If you mix equal molar quantities of the following substances, how many will produce an acidic solution?

Set 1: NaOH + HCl	Set 2: NaOH + HNO_3
Set 3: NH_3 + HCl	Set 4: NaOH + CH_3CO_2H

   a) four   b) three   c) two   d) one   e) zero (none are acidic)

17. If you mix equal molar quantities of the following substances, how many will produce a neutral solution?

Set 1: NaOH + HCl	Set 2: NaOH + HNO_3
Set 3: NH_3 + HCl	Set 4: NaOH + CH_3CO_2H

   a) four   b) three   c) two   d) one   e) zero (none are neutral)

	Acid	K_a	Conj. Base	K_b
Hydrofluoric Acid	HF	7.2×10^{-4}	F^-	1.4×10^{-11}
Nitrous Acid	HNO_2	4.5×10^{-4}	NO_2^-	2.2×10^{-11}
Acetic Acid	CH_3CO_2H	1.8×10^{-5}	$CH_3CO_2^-$	5.6×10^{-10}
Ammonium ion	NH_4^+	5.6×10^{-10}	NH_3	1.8×10^{-5}
Hydrocyanic acid	HCN	4.0×10^{-10}	CN^-	2.5×10^{-5}

18. If you mix equal molar quantities of the following substances, how many will produce a
    basic solution?

    Set 1: NaOH + HBr   Set 2: NaOH + HNO_3
    Set 3: NH_3 + HF   Set 4: NaOH + CH_3CO_2H

    a) four   b) three   c) two   d) one   e) zero (none are basic)

19. If you mix 100. mL of 0.11 M HCl with 50.0 mL of 0.22 M NH_3, what is the pH of the
    resulting solution?

    a) 4.63   b) 5.64   c) 6.02   d) 8.39   e) 9.37

20. If you mix 250. mL of 0.24 M HF with 75.0 mL of 0.80 M NaOH, what is the pH of the
    resulting solution?

    a) 5.42   b) 5.79   c) 6.24   d) 7.53   e) 8.21

21. If you mix 50.0 mL of 0.34 M HCN with 100. mL of 0.17 M KOH, what is the pH of the
    resulting solution?

    a) 11.23   b) 10.81   c) 4.60   d) 3.18   e) 2.77

22. If you mix 125. mL of 0.50 M CH_3CO_2H with 75.0 mL of 0.83 M NaOH, what is the pH of
    the resulting solution?

    a) 2.37   b) 2.62   c) 4.21   d) 10.40   e) 11.37

23. If you mix equal molar quantities of NaOH and CH_3CO_2H, what are the major species
    present in the resulting solution?
    a) Na^+, $CH_3CO_2^-$, OH^-, and H_2O
    b) Na^+, $CH_3CO_2^-$, CH_3CO_2H, OH^-, and H_2O
    c) Na^+, CH_3CO_2H, OH^-, and H_2O
    d) Na^+, $CH_3CO_2^-$, H_3O^+, and H_2O
    e) Na^+, CH_3CO_2H, H_3O^+, and H_2O

	Acid	K_a	Conj. Base	K_b
Hydrofluoric Acid	HF	7.2×10^{-4}	F^-	1.4×10^{-11}
Nitrous Acid	HNO_2	4.5×10^{-4}	NO_2^-	2.2×10^{-11}
Acetic Acid	CH_3CO_2H	1.8×10^{-5}	$CH_3CO_2^-$	5.6×10^{-10}
Ammonium ion	NH_4^+	5.6×10^{-10}	NH_3	1.8×10^{-5}
Hydrocyanic acid	HCN	4.0×10^{-10}	CN^-	2.5×10^{-5}

24. If you mix equal molar quantities of CH_3CO_2H and NaOH, the resulting solution will be

a) acidic because a small amount of CH_3CO_2H is present.

b) acidic because a small amount of H_3O^+ is present.

c) basic because a small amount of OH^- is present.

d) basic because a small amount of CH_3CO_2H is present.

e) neutral.

25. If you mix equal molar quantities of KOH and HNO_3, the resulting solution will be

a) acidic because a small amount of KNO_3 is present.

b) acidic because a small amount of H_3O^+ is present.

c) basic because a small amount of OH^- is present.

d) basic because a small amount of KNO_3 is present.

e) neutral.

26. If you mix equal molar quantities of NH_3 and CH_3CO_2H, the resulting solution will be

a) acidic because K_a of NH_4^+ is greater than K_b of $CH_3CO_2^-$

b) acidic because K_a of NH_4^+ is greater than K_a of CH_3CO_2H

c) basic because K_b of NH_3 is greater than K_b of $CH_3CO_2^-$

d) basic because K_a of NH_4^+ is greater than K_b of $CH_3CO_2^-$

e) neutral because K_a of NH_4^+ equals K_b of $CH_3CO_2^-$

27. If you mix equal molar quantities of NH_3 and HF, the resulting solution will be

a) acidic because K_a of NH_4^+ is greater than K_b for F^-

b) acidic because K_a of NH_4^+ is greater than K_a for HF

c) basic because K_a of NH_4^+ is greater than K_b for F^-

d) basic because K_a of NH_4^+ is greater than K_b for F^-

e) neutral.

	Acid	K_a	Conj. Base	K_b
Hydrofluoric Acid	HF	7.2×10^{-4}	F^-	1.4×10^{-11}
Nitrous Acid	HNO_2	4.5×10^{-4}	NO_2^-	2.2×10^{-11}
Acetic Acid	CH_3CO_2H	1.8×10^{-5}	$CH_3CO_2^-$	5.6×10^{-10}
Ammonium ion	NH_4^+	5.6×10^{-10}	NH_3	1.8×10^{-5}
Hydrocyanic acid	HCN	4.0×10^{-10}	CN^-	2.5×10^{-5}

28. What effect will the addition of the reagent in each of the following have on the pH of the $Ca(OH)_2$ solution respectively?

        Flask 1: Addition of HCl to $Ca(OH)_2$(aq)
        Flask 2: Addition of $CaCl_2$ to $Ca(OH)_2$(aq)

a) increase, increase   b) decrease, decrease   c) decrease, no change

d) decrease, increase   e) no change, decrease

29. What effect will the addition of the reagent in each of the following have on the pH of the HCN solution respectively?

        Flask 1: Addition of KCN to HCN(aq)
        Flask 2: Addition of KCl to HCN(aq)

a) no change, increase   b) no change, decrease   c) increase, no change

d) decrease, no change   e) decrease, decrease

30. What effect will the addition of the reagent in each of the following have on the pH of the CH_3CO_2H solution respectively?

        Flask 1: Addition of $NaOH_3CO_2$ to CH_3CO_2H(aq)
        Flask 2: Addition of $Ca(CH_3CO_2)_2$ to CH_3CO_2H(aq)

a) no change, increase   b) no change, decrease   c) decrease, no change

d) decrease, decrease   e) increase, increase

31. What effect will the addition of the reagent in each of the following have on the pH of the NH_3 solution respectively?

        Flask 1: Addition of NH_4Cl to NH_3(aq)
        Flask 2: Addition of $CaCl_2$ to NH_3(aq)

a) increase, increase   b) decrease, decrease   c) decrease, no change

d) decrease, increase   e) no change, decrease

32. If you add 0.82 g $NaCH_3CO_2$ to 100. mL of a 0.10 M CH_3CO_2H solution, what is the new pH of the resulting solution? For CH_3CO_2H $K_a = 1.8 \times 10^{-5}$.

a) 1.73   b) 2.40   c) 3.20   d) 4.76   e) 5.74

	Acid	K_a	Conj. Base	K_b
Hydrofluoric Acid	HF	7.2×10^{-4}	F^-	1.4×10^{-11}
Nitrous Acid	HNO_2	4.5×10^{-4}	NO_2^-	2.2×10^{-11}
Acetic Acid	CH_3CO_2H	1.8×10^{-5}	$CH_3CO_2^-$	5.6×10^{-10}
Ammonium ion	NH_4^+	5.6×10^{-10}	NH_3	1.8×10^{-5}
Hydrocyanic acid	HCN	4.0×10^{-10}	CN^-	2.5×10^{-5}

33. If you add 0.35 g NaF to 150. mL of a 0.30 M HF solution, what is the new pH of the resulting solution? For HF, $K_a = 7.2 \times 10^{-4}$

    a) 2.41   b) 3.15   c) 3.90   d) 4.21   e) 5.55

34. If you add 0.42 g NH_4Cl to 200. mL of a 0.34 M NH_3 solution, what is the new pH of the resulting solution? For NH_3, $K_b = 1.8 \times 10^{-5}$

    a) 3.81   b) 4.74   c) 7.92   d) 9.25   e) 10.20

35. If you add 0.50 g KNO_2 to 175 mL of a 0.24 M HNO_2 solution, what is the new pH of the resulting solution? For HNO_2 Ka= 4.4×10^{-4}

    a) 1.81   b) 2.50   c) 3.35   d) 4.72   e) 5.87

36. If you add 1.0 mL of 10.0 M HCl to 500. mL of a 0.10 M NH_3 solution, what is the pH of the resulting solution?

    a) 4.05   b) 5.80   c) 8.21   d) 9.25   e) 9.95

37. If you add 1.0 mL of 5.0 M NaOH to 250. mL of a 0.35 M CH_3CO_2H solution, what is the pH of the resulting solution?

    a) 1.54   b) 2.89   c) 3.50   d) 4.74   e) 5.36

38. If you add 5.0 mL of 0.20 M HCl to 250. mL of a 0.42 M $NaCH_3CO_2$ solution, what is the pH of the resulting solution?

    a) 6.00   b) 7.00   c) 8.00   d) 10.00   e) 12.00

39. If you add 10.0 mL of 0.23 M NH_3 to 150. mL of a 3.17 M NH_4Cl solution, what is the pH of the resulting solution?

    a) 4.50   b) 5.26   c) 6.90   d) 7.85   e) 8.32

	Acid	K_a	Conj. Base	K_b
Hydrofluoric Acid	HF	7.2×10^{-4}	F^-	1.4×10^{-11}
Nitrous Acid	HNO_2	4.5×10^{-4}	NO_2^-	2.2×10^{-11}
Acetic Acid	CH_3CO_2H	1.8×10^{-5}	$CH_3CO_2^-$	5.6×10^{-10}
Ammonium ion	NH_4^+	5.6×10^{-10}	NH_3	1.8×10^{-5}
Hydrocyanic acid	HCN	4.0×10^{-10}	CN^-	2.5×10^{-5}

40. If you add 1.0 mL of 2.0 M HCl to 500. mL of a 0.44 M $NaNO_2$ solution, what is the pH of the resulting solution?

    a) 1.75   b) 7.25   c) 9.25   d) 10.75   e) 12.25

41. At the neutralization point of the titration of an acid with base, what condition is met?

    a) volume of base added from buret equals volume acid in reaction flask.

    b) molarity of base from the buret equals molarity of acid in reaction flask

    c) moles of base added from the buret equals moles of acid in the reaction flask

    d) % ionization of base added from the buret equals % ionization of the acid in flask.

    e) all of the above conditions are met

42. The salt produced by the reaction of an equal number of moles of KOH and HNO_3 will react with water to give a solution which is

    a) acidic   b) basic   c) neutral
    d) non-ionic   e) impossible to determine

43. If CH_3CO_2H reacts with an equal number of moles of NaOH then the pH of the resulting solution is approximately.

    a) 1   b) 3   c) 7   d) 9   e) 13

44. What is the pH at the equivalence point in the titration of 10.0 mL of 0.32 M CH_3CO_2H with 0.213 M NaOH?

    a) 7.00   b) 3.20   c) 11.18   d) 5.07   e) 8.93

45. What is the pH at the equivalence point in the titration of 10.0 mL of 0.16 M NH_3 with 0.064 M HCl?

    a) 4.75   b) 5.30   c) 7.00   d) 8.70   e) 9.25

		Acid	Ka	Conj. Base	Kb
Hydrofluoric Acid	HF	7.2×10^{-4}	F^-	1.4×10^{-11}
Nitrous Acid	HNO_2	4.5×10^{-4}	NO_2^-	2.2×10^{-11}
Acetic Acid	CH_3CO_2H	1.8×10^{-5}	$CH_3CO_2^-$	5.6×10^{-10}
Ammonium ion	NH_4^+	5.6×10^{-10}	NH_3	1.8×10^{-5}
Hydrocyanic acid	HCN	4.0×10^{-10}	CN^-	2.5×10^{-5}

46. What is the pH at the equivalence point in the titration of 20.0 mL of 0.64 M HF with 0.32 M NaOH?

a) 7.00   b) 5.07   c) 8.93   d) 3.25   e) 10.75

47. What is the pH and molar concentration the Na^+ and NO_3^-, at the equivalence point in the titration of 25.0 mL 0.48 M HNO_3 with 0.60 M NaOH?

a) pH = 5 $[Na^+] = 0.012$ $[NO_3^-] = 0.012$

b) pH = 5 $[Na^+] = 0.045$ $[NO_3^-] = 0.012$

c) pH = 7 $[Na^+] = 0.12$ $[NO_3^-] = 0.12$

d) pH = 7 $[Na^+] = 0.45$ $[NO_3^-] = 0.45$

e) pH = 9 $[Na^+] = 0.012$ $[NO_3^-] = 0.45$

48. Which of the following combinations will NOT produce a buffer solution?

a) NaCl and CH_3CO_2H b) NH_4Cl and NH_3 c) KCN and HCN

d) $NaHCO_3$ and H_2CO_3 e) $NaCH_3CO_2$ and CH_3CO_2H

49. Which of the following combinations can produce a buffer solution?

a) NaCl and CH_3CO_2H b) NH_4Cl and HNO_3 c) KCl and HCN

d) $NaCH_3CO_2$ and CH_3CO_2H e) $NaNO_3$ and HF

50. A buffer solution of pH=5.30 can be prepared by dissolving acetic acid and sodium acetate in water. How many moles of sodium acetate must be added to 1.0 L of 0.25 M acetic acid to prepare the buffer?

a) 0.28 mol   b) 0.47 mol   c) 0.90 mol   d) 1.8 mol   e) 3.6 mol

51. A buffer solution of pH=9.35 can be prepared by dissolving ammonia and ammonium chloride in water. How many moles of ammonium chloride must be added to 1.0 L of 0.50 M ammonia to prepare the buffer?

a) 0.40 mol   b) 0.80 mol   c) 1.2 mol   d) 2.5 mol   e) 4.5 mol

Chapter 18 : Answers to Multiple Choice

11. a	21. a	31. c
12. b	22. e	32. d
13. e	23. b	33. a
14. a	24. c	34. e
15. c	25. e	35. b
16. d	26. e	36. e
17. c	27. a	37. c
18. d	28. b	38. c
19. c	29. c	39. c
20. e	30. e	40. b

41. c	51. a
42. c	
43. d	
44. e	
45. b	
46. c	
47. d	
48. a	
49. d	
50. c	

Chapter 19
Precipitation Reactions

Section A: Free Response

1. According to the general solubility rules, silver chloride is "insoluble" and a handbook lists its solubility product as 1.8×10^{-10}. However, when a chemist treated a solution of $AgNO_3$ with a solution of HCl a precipitate did NOT form. Explain how this could have happened. (Assume the bottles were labeled correctly.)

2. Calculate Q, the reaction quotient, for the mixture obtained when solutions X and Y are mixed. Based on your value of Q, determine if a precipitate will form.
   K_{sp} for $PbSO_4 = 1.3 \times 10^{-8}$
   Solution X: 200.0 mL of 3.8×10^{-3} M $Pb(NO_3)_2$
   Solution Y: 300.0 mL of 4.1×10^{-4} M H_2SO_4

3. What is the solubility product constant for AgOH(silver hydroxide) if a saturated solution of AgOH in water has a pH = 10.15?

4. The K_{sp} value for some salts does not adequately predict the solubility in water because other simultaneous equilibria occur with the ions from the salt. Consider the salts, MnS, $Mn(OH)_2$ and $PbCl_2$. Which one has solubility in water greater than the Ksp value suggests? Explain. using appropriate equations.

5. Of the three compounds $BaCO_3$, $Bi(OH)_3$, and AgCl, how many will be more soluble in acid than in pure water. Write appropriate equations to explain your answer.

6. How does the solubility of $CaCO_3$ in pure water compare with the solubility of $CaCO_3$ in a solution which also contains $CaCl_2$? How does the solubility of $CaCO_3$ in pure water compare with the solubility of $CaCO_3$ in a solution which also contains NaCl? Use equations to explain your answer.

7. List the compounds silver carbonate, silver chloride, and silver bromide in an order of increasing molar solubility.

Ag_2CO_3 ; $K_{sp} = 8.1 \times 10^{-12}$ $AgCl$; $K_{sp} = 1.8 \times 10^{-10}$ $AgBr$; $K_{sp} = 3.3 \times 10^{-13}$

8. If NaCl is added to a solution that is 0.010 M in $Ag+$, $Cu+$, and $Au+$, which compound will precipitate first.

$AgCl$; $K_{sp} = 1.8 \times 10^{-10}$ $CuCl$; $K_{sp} = 1.9 \times 10^{-7}$ $AuCl$; $K_{sp} = 2.0 \times 10^{-13}$

9. For $PbCl_2$, $Ksp = 1.7 \times 10^{-5}$. How many grams of NaCl must be added to 100. mL of 0.0018 M $Pb(NO_3)_2$ to begin precipitating $PbCl_2$?

10. The formation constant for $Ag(NH)_2{}^+$ is 1.6×10^7 and the solubility product constant for AgCl is 1.8×10^{-8}. Use this information and appropriate equations to explain the following student's observations:

> A few drops of concentration sodium chloride was added to a silver nitrate solution   forming a white precipitate. Upon addition of concentrated ammonia, the precipitate dissolved.

Key Concepts for Free Response

1. The solutions may have been so dilute that the AgCl could not precipitate. If the concentrations are not at a certain level (K_{sp} has not been exceeded) then no precipitate will form.

2. $PbSO_4 (s) \rightleftharpoons Pb^{2+}(aq) + SO_4^{2-}(aq)$

   $Q = 3.74 \times 10^{-7}$ which is greater than Ksp. Yes a precipitate forms.

3. K_{sp} for AgOH $= 2.0 \times 10^{-8}$

4. MnS because sulfides can react with water shifting the equilibrium to the right

       $MnS (s) \rightleftharpoons Mn^{2+}(aq) + S^{2-}(aq)$

       $S^{2-} (aq) + H_2O(l) \rightleftharpoons HS^- (aq) + OH^- (aq)$

   $Mn(OH)_2$ and $PbCl_2$ would have this additional factor to increase solubility.

5. $BaCO_3$ and $Bi(OH)_3$ will be more soluble in acid than in water. AgCl will not be more soluble in acid. $BaCO_3$ will react with acid according to the following equations. The formation of CO2 drive the reaction to completion, dissolving all the $BaCO_3$.

       $BaCO_3 (s) \rightleftharpoons Ba^{2+}(aq) + CO_3^{2-}(aq)$

       $CO_3^{2-} (aq) + 2 H_3O^+(aq) \text{\AE} H_2CO_3(aq) + 2 H_2O(l)$
       $H_2CO_3(aq) \dashrightarrow H_2O(l) + CO_2(g)$

   Similarly the $Bi(OH)_3$ dissolves because the acid solution reacts with the OH^-.

       $Bi(OH)_3 (s) \rightleftharpoons Bi^{3+}(aq) + 3 OH^-(aq)$

       $OH^- (aq) + H_3O^+(aq) \dashrightarrow 2 H_2O(l)$

6. $CaCO_3$ is less soluble in a solution that contains $CaCl_2$ than in water. The common ion, Ca^{2+} will shift the equilibrium to the left and decrease the solubility according to the equation

       $CaCO_3 (s) \rightleftharpoons Ca^{2+}(aq) + CO_3^{2-}(aq)$

   $CaCO_3$ should have the same solubility in a solution that contains NaCl as in water. There are no common ions or other equilibria to change the solubility.

7. The order of solubility is $AgBr < AgCl < Ag_2CO_3$

8. AuCl because the added Cl- will cause its Ksp to be exceeded before the other compounds.

9. 0.60 grams of NaCl will cause $PbCl_2$ to begin precipitating.

10. AgCl is formed when the sodium chloride is added. K_{sp} is a small number so a few drops caused K_{sp} to be exceeded. The solution then contained

        $AgCl (s) \rightleftharpoons Ag^+(aq) + Cl^-(aq)$

    When ammonia was added the solid dissolved according to the equation

        $Ag^+(aq) + NH_3 (aq) \rightleftharpoons Ag(NH_3)_2^+(aq)$

    The large value of the formation constant indicates that the equilibrium lies far to the right so it removes the Ag^+ from the precipitation equilibrium and the AgCl dissolves.

Section B: Multiple Choice

11. Which of the following is the solubility product constant for PbI_2?

    a) $K_{sp} = [Pb^{2+}] [2I^-]^2$ b) $K_{sp} = [Pb^{2+}] [2I^-]$

    c) $K_{sp} = [Pb^{2+}]^2 [I^-]$ d) $K_{sp} = [Pb^{2+}] [I_2^-]^2$

    e) $K_{sp} = [Pb^{2+}] [I^-]^2$

12. Which of the following is the solubility product constant for $Ca_3(PO_4)_2$?

    a) $K_{sp} = [Ca^{2+}] [PO_4^{3-}]^2$ b) $K_{sp} = [Ca^{2+}]^2 [PO_4^{3-}]^3$

    c) $K_{sp} = [Ca^{2+}]^3 [PO_4^{3-}]^2$ d) $K_{sp} = [Ca^{2+}] [PO_4^{3-}]^2$

    e) $K_{sp} = [Ca^{2+}]^6 [PO_4^{3-}]^6$

13. Which of the following is the solubility product constant for iron (III) sulfide, Fe_2S_3 ?

    a) $K_{sp} = [Fe^{2+}] [S^{3-}]$ b) $K_{sp} = [Fe^{3+}] [S^{2-}]^3$

    c) $K_{sp} = [Fe^{2+}]^2 [S^{3-}]^3$ d) $K_{sp} = [Fe^{3+}]^2 [S^{2-}]^3$

    e) $K_{sp} = [Fe^{2+}] [S^{2-}]^3$

14. Which of the following is the solubility product constant for $Mn(OH)_2$?

    a) $K_{sp} = [Mn^{2+}] [OH^-]^2$ b) $K_{sp} = [Mn^{2+}] [2OH^-]^2$

    c) $K_{sp} = [Mn^{2+}]^2 [OH^-]^2$ d) $K_{sp} = [Mn^{2+}]^2 [OH^-]$

    e) $K_{sp} = [Mn^{2+}]^2 [2OH^-]^2$

15. The solubility of $FeCO_3$ is 5.9×10^{-6} mol/L . What is K_{sp} for $FeCO_3$?

    a) 5.9×10^{-6} b) 1.2×10^{-21} c) 3.5×10^{-11}

    d) 2.8×10^{-10} e) 1.3×10^{-14}

16. The solubility of HgS is 5.5×10^{-27}. What is K_{sp} for $Zn(CN)_2$?

    a) 4.0×10^{-3} b) 8.2×10^{-4} c) 1.3×10^{-13}

    d) 7.4×10^{-14} e) 3.0×10^{-53}

17. The solubility of $PbBr_2$ is 0.0116 M. What is K_{sp} for $PbBr_2$?

    a) 1.6×10^{-6} b) 6.3×10^{-6} c) 1.3×10^{-4}

    d) 3.1×10^{-6} e) 1.1×10^{-1}

18. The solubility of Ag_2SO_4 0.0162 M . What is K_{sp} for Ag_2SO_4?
    a) 1.7×10^{-5} b) 6.1×10^{-3} c) 2.6×10^{-4}
    d) 1.4×10^{-4} e) 5.2×10^{-4}

19. Which of the following expressions describes the relationship between the solubility product, Ksp, and the solubility, s, of MgF_2?
    a) $K_{sp} = 2s$ b) $K_{sp} = s^2$ c) $K_{sp} = 2s^3$
    d) $K_{sp} = 4s^2$ e) $K_{sp} = 4s^3$

20. Which of the following compounds has the highest molar solubility?
    a) $AgCl$; $K_{sp} = 1.8 \times 10^{-10}$ b) $CuBr$; $K_{sp} = 5.3 \times 10^{-9}$
    c) $AgBr$; $K_{sp} = 3.3 \times 10^{-13}$ d) CuI ; $K_{sp} = 5.1 \times 10^{-12}$
    e) $CuCl$; $K_{sp} = 1.9 \times 10^{-7}$

21. Which of the following compounds has the highest molar solubility?
    a) $BaSO_4$; $K_{sp} = 1.1 \times 10^{-10}$ b) $FeCO_3$; $K_{sp} = 3.5 \times 10^{-11}$
    c) $PbSO_4$; $K_{sp} = 1.8 \times 10^{-8}$ d) $SrCO_3$; $K_{sp} = 9.4 \times 10^{-10}$
    e) $ZnCO_3$; $K_{sp} = 1.5 \times 10^{-11}$

22. Which of the following compounds has the lowest molar solubility?
    a) $BaSO_4$; $K_{sp} = 1.1 \times 10^{-10}$ b) $FeCO_3$; $K_{sp} = 3.5 \times 10^{-11}$
    c) $PbSO_4$; $K_{sp} = 1.8 \times 10^{-8}$ d) $SrCO_3$; $K_{sp} = 9.4 \times 10^{-10}$
    e) $ZnCO_3$; $K_{sp} = 1.5 \times 10^{-11}$

23. Rank the compounds from lowest to highest molar solubility.
    $$FeCO_3; \quad K_{sp} = 3.5 \times 10^{-11}$$
    $$BaSO_4 ; \quad K_{sp} = 1.1 \times 10^{-10}$$
    $$ZnCO_3; \quad K_{sp} = 1.5 \times 10^{-11}$$
    a) $ZnCO_3 > BaSO_4 > FeCO_3$ b) $FeCO_3 > ZnCO_3 > BaSO_4$
    c) $ZnCO_3 > FeCO_3 > BaSO_4$ d) $BaSO_4 > ZnCO_3 > FeCO_3$
    e) $BaSO_4 > FeCO_3 > ZnCO_3$

24. Which of the following compounds has the highest molar solubility?
    a) $PbCO_3$; $K_{sp} = 1.5 \times 10^{-13}$ b) PbS; $K_{sp} = 8.4 \times 10^{-28}$
    c) $PbSO_4$; $K_{sp} = 1.8 \times 10^{-8}$ d) $PbCrO_4$; $K_{sp} = 1.8 \times 10^{-14}$
    e) PbI_2; $K_{sp} = 8.7 \times 10^{-9}$

25. Which of the following will give a saturated solution with the highest concentration of
    iodide ion, I^-?
    a) CuI ; $K_{sp} = 5.1 \times 10^{-12}$ b) AuI; $K_{sp} = 1.0 \times 10^{-46}$
    c) AgI; $K_{sp} = 1.5 \times 10^{-16}$ d) BiI_3 ; $K_{sp} = 8.1 \times 10^{-19}$
    e) AuI_3; $K_{sp} = 1.6 \times 10^{-23}$

26. What is the concentration of SO_4^{2-} in a saturated solution of $BaSO_4$ if Ksp= 1.1×10^{-10}?
    a) 1.1×10^{-10} M   b) 5.5×10^{-11} M   c) 5.0×10^{-5} M
    d) 1.0×10^{-5} M   e) 9.5×10^{-4} M

27. What is the concentration of CrO_4^{2-} in a saturated solution of $PbCrO_4$ if
    Ksp= 1.8×10^{-14}?
    a) 1.3×10^{-7} M   b) 7.5×10^{-6} M   c) 1.8×10^{-4} M
    d) 1.3×10^{-4} M   e) 5.1×10^{-3} M

28. What is the concentration of F^- in a saturated solution of BaF_2 if Ksp= 1.7×10^{-6}?
    a) 7.5×10^{-3} M   b) 8.2×10^{-4} M   c) 1.5×10^{-2} M
    d) 4.3×10^{-7} M   e) 1.5×10^{-6} M

29. If 15.0 mg of AgCl is placed in 100.0 mL water, what is the chloride ion concentration
    the solution? $K_{sp} = 1.8 \times 10^{-10}$
    a) 4.8×10^{-3} M   b) 9.6×10^{-3} M   c) 1.0×10^{-4} M
    d) 1.34×10^{-5} M   e) 1.8×10^{-10} M

30. If 15.0 mg of $BaCO_3$ is placed in 250. mL water, what is the barium ion concentration in
    the solution? $K_{sp} = 8.1 \times 10^{-9}$
    a) 3.0×10^{-3} M   b) 6.0×10^{-4} M   c) 9.0×10^{-5} M
    d) 1.5×10^{-6} M   e) 7.6×10^{-6} M

31. If 2.00 mg of CaF_2 is placed in 500. mL water, what is the calcium ion concentration in the solution? $K_{sp} = 3.9 \times 10^{-11}$
    a) 2.6×10^{-5} M   b) 5.1×10^{-5} M   c) 2.1×10^{-4} M
    d) 4.2×10^{-8} M   e) 1.2×10^{-6} M

32. If 2.00 mg of CaF_2 is placed in 500. mL water, what is the fluoride ion concentration in the solution? $K_{sp} = 3.9 \times 10^{-11}$
    a) 1.0×10^{-4} M   b) 2.1×10^{-4} M   c) 5.1×10^{-5} M
    d) 2.0×10^{-8} M   e) 2.5×10^{-3} M

33. Calculate the equilibrium constant for the reaction:
$$CuCl(s) + I^-(aq) \rightleftharpoons CuI(s) + Cl^-(aq)$$
    $CuCl;\ K_{sp} = 1.9 \times 10^{-7}$ $CuI;\ K_{sp} = 5.1 \times 10^{-12}$
    a) 8.4×10^{-2} b) 2.3×10^{-6} c) 3.7×10^4
    d) 4.4×10^{17} e) 9.7×10^{-19}

34. Calculate the equilibrium constant for the reaction:
$$CdS(s) + Zn^{2+}(aq) \rightleftharpoons ZnS(s) + Cd^{2+}(aq)$$
    $CdS;\ K_{sp} = 3.6 \times 10^{-29}$ $CuI;\ K_{sp} = 1.1 \times 10^{-21}$
    a) 3.3×10^{-8} b) 2.7×10^{-4} c) 4.2×10^5
    d) 2.5×10^{49} e) 3.1×10^7

35. For AgI, $K_{sp} = 1.5 \times 10^{-16}$. If you mix 500 mL of 1×10^{-8} M $AgNO_3$ and 500 mL of 1×10^{-8}M NaI, what will be observed?
    a) a precipitate forms because $Q_{sp} > K_{sp}$
    b) a precipitate forms because $Q_{sp} < K_{sp}$
    c) no precipitate forms because $Q_{sp} = K_{sp}$
    d) no precipitate forms because $Q_{sp} < K_{sp}$
    e) no precipitate forms because $Q_{sp} > K_{sp}$

36. For $BaSO_4$, $K_{sp} = 1.1 \times 10^{-10}$. If you mix 200. mL of 1.0×10^{-4} M $Ba(NO_3)_2$ and 500. mL of 800. x 10^{-4} M H_2SO_4, what will be observed?
    a) a precipitate forms because $Q_{sp} > K_{sp}$
    b) a precipitate forms because $Q_{sp} < K_{sp}$
    c) no precipitate forms because $Q_{sp} = K_{sp}$
    d) no precipitate forms because $Q_{sp} < K_{sp}$
    e) no precipitate forms because $Q_{sp} < K_{sp}$

37. For MgF_2, $K_{sp} = 6.4 \times 10^{-9}$. If you mix 400. mL of 1×10^{-4} M $Mg(NO_3)_2$ and 500. mL of 1.00×10^{-4} M NaF, what will be observed?
    a) a precipitate forms because $Q_{sp} > K_{sp}$
    b) a precipitate forms because $Q_{sp} < K_{sp}$
    c) no precipitate forms because $Q_{sp} = K_{sp}$
    d) no precipitate forms because $Q_{sp} < K_{sp}$
    e) no precipitate forms because $Q_{sp} < K_{sp}$

38. For $ZnCO_3$, $K_{sp} = 1.5 \times 10^{-11}$. If you mix 250. mL of 2.0×10^{-3} M $ZnCl_2$ and 750. mL of 4.0×10^{-8} M $CaCO_3$, what will be observed?
    a) a precipitate forms because $Q_{sp} > K_{sp}$
    b) a precipitate forms because $Q_{sp} < K_{sp}$
    c) no precipitate forms because $Q_{sp} = K_{sp}$
    d) no precipitate forms because $Q_{sp} < K_{sp}$
    e) no precipitate forms because $Q_{sp} < K_{sp}$

39. A saturated solution of $Ca(OH)_2$ has a pH = 12.40. What is K_{sp} for $Ca(OH)_2$?
    a) 2.0×10^{-6} b) 1.3×10^{-2} c) 7.9×10^{-6}
    d) 4.0×10^{-13} e) 2.5×10^{-2}

40. A saturated solution of cobalt hydroxide, $Co(OH)_2$ has a pH = 8.90. What is K_{sp} for $Co(OH)_2$?
    a) 2.1×10^{-27} b) 5.0×10^{-16} c) 4.0×10^{-12}
    d) 2.5×10^{-16} e) 1.3×10^{-2}

41. The solubility of salts can be affected by other equilibria. How is the solubility of $ZnCO_3$ be changed by the following reaction?

$$CO_3^{2-}(aq) + H_2O(l) \rightleftharpoons HCO_3^-(aq) + OH^-(aq)$$

a) the solubility is increased because the CO_3^{2-} can undergo further reaction.

b) the solubility is decreased because the CO_3^{2-} can undergo further reaction.

c) the solubility is increased because a gas is formed

d) the solubility is decreased because a gas is formed

e) the solubility stays the same because as more $ZnCO_3$ is dissolved, more is precipitated
   simultaneously.

42. The solubility of salts can be affected by other equilibria. How is the solubility of PbS be changed by the following reaction?

$$S^{2-}(aq) + H_2O(l) \rightleftharpoons HS^-(aq) + OH^-(aq)$$

a) the solubility is increased because a gas is formed

b) the solubility is decreased because a gas is formed

c) the solubility stays the same because the solution is saturated

d) the solubility is increased because the S^{2-} can undergo further reaction.

e) the solubility is decreased because the S^{2-} can undergo further reaction.

43. For $BaSO_4$, $K_{sp} = 1.1 \times 10^{-10}$. What is the solubility of $BaSO_4$ in a solution which is 0.018 M in Na_2SO_4?

a) 0.018 mol/L   b) 1.1×10^{-5} mol/L   c) 1.1×10^{-10} mol/L

d) 6.1×10^{-9} mol/L   e) 7.8×10^{-5} mol/L

44. For AgI, $K_{sp} = 8.3 \times 10^{-17}$. What is the solubility of AgI in a solution which is 5.1×10^{-4} M in $AgNO_3$?

a) 8.3×10^{-11} mol/L   b) 5.1×10^{-2} mol/L   c) 1.1×10^{-5} mol/L

d) 4.2×10^{-20} mol/L   e) 1.6×10^{-13} mol/L

45. For $BaSO_4$, $K_{sp} = 1.1 \times 10^{-10}$. If 1.00 g of $BaSO_4$ is placed in 1.00 L of pure water at 25°C, how much of will dissolve?

a) all of it   b) 0.0025 g   c) 4.0×10^{-2} g

d) 1.5×10^{-4} g   e) 0.35 g

46. For $NiCO_3$, $K_{sp} = 6.6 \times 10^{-9}$. If 0.050 g of $NiCO_3$ is placed in 1.00 L of pure water at 25°C, how much of will dissolve?

a) all of it   b) 0.044 g   c) 0.037 g

d) 0.025 g   e) 0.010 g

47. For CaF_2, $K_{sp} = 3.9 \times 10^{-11}$. If 0.0080 g of CaF_2 is placed in 1.00 L of pure water at 25°C, how much of will dissolve?

a) all of it   b) 0.00022 g   c) 0.0040 g

d) 0.0025   e) 0.0079 g

48. For Ag_2SO_4, $K_{sp} = 1.4 \times 10^{-5}$. If 1.00 g of Ag_2SO_4 is placed in 2.00 L of pure water at 25°C, how much of will dissolve?

a) all of it   b) 0.94 g   c) 0.47 g   d) 0.18 g   e) 0.014 g

49. For Ag_2SO_4, $K_{sp} = 1.4 \times 10^{-5}$. How many grams of Na_2SO_4 must be added to 100. mL of 0.022 M $AgNO_3$ to begin to see a precipitate form?

a) 0.37 g   b) 3.7 g   c) 0.029   d) 0.17 g   e) 1.4 g

50. For thallium bromide, TlBr, $K_{sp} = 3.4 \times 10^{-6}$. How many grams of KBr must be added to 100. mL of 0.0055 M $TlNO_3$ to begin to see a precipitate form?

a) 0.0041 g   b) 0.062 g   c) 0.73 g   d) 3.4 g   e) 1.8 g

51. For $PbCl_2$, $K_{sp} = 1.7 \times 10^{-5}$. How many grams of NaCl must be added to 200. mL of 0.16 M $Pb(NO_3)_2$ to begin to see a precipitate form?

a) 0.021 g   b) 2.34 g   c) 0.85 g   d) 0.43 g   e) 0.12 g

52. For $CaCrO_4$, $K_{sp} = 7.1 \times 10^{-4}$. How many grams of Na2CrO4 must be added to 200. mL of 0.250 M $Ca(NO_3)_2$ to begin to see a precipitate form?

a) 0.046 g   b) 0.092 g   c) 0.34 g   d) 1.8 g   e) 2.1 g

Chapter 19 : Answers to Multiple Choice

11. e	21. c	31. b
12. c	22. b	32. a
13. d	23. c	33. c
14. a	24. e	34. a
15. c	25. d	35. d
16. e	26. d	36. a
17. b	27. a	37. d
18. a	28. c	38. a
19. e	29. d	39. c
20. e	30. c	40. d

41. a	51. e
42. d	52. b
43. d	
44. e	
45. b	
46. e	
47. a	
48. a	
49. d	
50. c	

Chapter 20
Principles of Reactivity: Entropy an Free Energy

Section A: Free Response

1. Why is the entropy of vaporization of water greater than that of a typical liquid?

2. Why is the entropy of a gas greater than the entropy of a liquid?

3. Does the following reaction favor products or reactants? Why? If the reaction favors reactants, at what temperature does the products become favored? Use values of ΔG^o.

$$HNO_3(l) + H_2O(l) \rightarrow NH_3(g) + 2\,O_2(g)$$

4. Predict the sign of the entropy change for the following reactions and explain. Which would you expect to have the greatest entropy change.
   (a) $C_8H_{16}(g) + H_2(g) \rightarrow C_8H_{18}(g)$
   (b) $C_2H_5OH(l) + 3\,O_2(g) \rightarrow 2\,CO_2(g) + 3\,H_2O(g)$
   (c) $Na_2CO_3(aq) + 2\,HCl(aq) \rightarrow 2\,NaCl(aq) + H_2O(l) + CO_2(g)$

5. Using values of ΔH_f^o and S^o, calculate ΔG^o_{rxn} for the following reaction. Is the reaction spontaneous? if so, comment on whether it is entropy or enthalpy driven.

$$NH_3(g) + HCl(g) \rightarrow NH_4Cl(s)$$

ΔH(kJ/mol)	-46.1	-92.3	-314
S^o(J/K · mol)	192.5	187	94.6

6. Calculate the equilibrium constant for a given reaction in which $\Delta G^o = -52.6$ kJ. According to the equilibrium constant, will the reaction favor products or reactants? Comment on the connection between the sign of ΔG^o and the equilibrium constant.

7. By looking at the state or formula of each of the following pairs of compounds, predict which one will have the greatest entropy at the same temperature.
   (a) C(diamond) or C(graphite)
   (b) $NaCl(s)$ or $CaCl_2(s)$
   (c) $NaNO_3(aq)$ or $NaNO_3(s)$
   (d) $Fe(s)$ or $Br_2(l)$

8. Calculate the change in entropy for the decomposition of methanol to give methane and oxygen under standard conditions. Do you agree with the sign? Why or why not? Is the reaction spontaneous or not spontaneous? If it is not, at what temperature does it become so?

	$S°$(J/mol · K)	$\Delta H_f°$(kJ/mol)
CH_4(g)	186.3	-75.0
CH_3OH(l)	126.8	-237
O_2(g)	205.1	

9. State and explain the three laws of thermodynamics.

10. What does the term *spontaneous* mean when referring to a chemical reaction such as the
$$Mg(s) + CO(g) \rightarrow MgO(s) + C(graphite) ?$$

Key Concepts of Free Response

1. Water forms two hydrogen bonds. Therefore, it takes more energy to vaporize the liquid, which in turn makes the entropy of vaporization higher from the equation $\Delta S_{vap} = \Delta H_{vap}/T$.

2. Gaseous molecules are more scattered. There is no pattern to a gas; therefore, the randomness is increased.

3. The reaction is reactant favored because the free energy is positive (301.3 kJ). The reaction would become spontaneous at temperatures above 12350 K.

4. Reaction (a) will have a negative entropy sign. Two gaseous substances are combining to form one gaseous product, therefore randomness is decreased. Reaction (b) will have a positive entropy.

   $\Delta n_{gas} = 2$. Five moles of gas is created from 3 moles of gas and one mole of liquid, therefore randomness is increased. Reaction (c) will also have a positive entropy. One mole of gas is created and four moles of products are created from 3 moles of aqueous substances, therefore the randomness has increased. The reaction that will have the greatest entropy will be reaction b because $\Delta n_{gas} = 2$, it has the greatest increase in randomness.

5. The reaction is spontaneous, $\Delta G = -91.2$ kJ. It is enthalpy driven because the change in entropy is negative, therefore the enthalpy change is a large negative number to give the free energy a negative value overall.

6. $K = 1.66 \times 10^9$ and will favor the products. If ΔG^o is negative then the equilibrium constant will be greater than one, but if it is positive then the equilibrium constant will be less than one. This fact is written into the equation, making the free energy equal to the negative ln K.

   Also, if ΔG^o is negative then the reaction is spontaneous, therefore the equilibrium constant must be greater than one.

7. (a) graphite, (b) $CaCl_2(s)$, (c) $NaNO_3(aq)$, (d) $Br_2(l)$

8. The change in entropy is +162.1 J/K which is expected since 1.5 moles of gas is evolved from 1 mole of liquid. The reaction however, is not spontaneous, the free energy is +114 kJ. The reaction will become spontaneous at 1006 K.

9. *First law*: The total energy of the universe is a constant.
   Second law: The total entropy of the universe is always increasing.
   Third law: The entropy of a pure, perfectly formed crystalline substance at absolute zero is zero.

10. A spontaneous chemical reaction means simply that the reaction will proceed as written and the products will be favored. In this case, magnesium oxide and carbon will be formed.

Section B: Multiple Choice

11. If the reaction $A + B \rightarrow C$ has an equilibrium constant less than 1. Which of the following statements are true?

    a) the reaction is non-spontaneous   b) the reaction is spontaneous

    c) the reaction will not occur   d) the reaction will happen instantly

    e) the reaction will explode

12. The disorder of a system is measured by

    a) enthalpy   b) Gibbs free energy   c) entropy

    d) heat of vaporization   e) equilibrium constant

13. The enthalpy of vaporization of methanol (CH_3OH) is 35.3 kJ/mol at the boiling point of 64.2 °C. Calculate the entropy change for methanol going from a liquid to vapor.

    a) -104.7 J/K·mol   b) -551.0 J/K·mol   c) +551.0 J/K·mol

    d) +600.0 J/K·mol   e) +104.7 J/K·mol

14. The change in entropy for the vaporization of CCl_4 is +85.7 J/K·mol and the boiling temperature is 77 °C. What is the heat of vaporization?

    a) +245 kJ/mol   b) +30.0 kJ/mol   c)+4.08 kJ/mol

    d) -245 kJ/mol   e)+1,110 kJ/mol

Table 20-1. Standard Entropy Values.

Compound	ΔS(J/K·mol)	Compound	ΔS(J/K·mol)
C(graph)	5.740		
C(diamond)	2.38	N(g)	153.3
H_2O(l)	69.9	N_2(g)	191.6
C_2H_6(g)	229.6	NO(g)	210.8
CH_3CH_2OH(l)	160.7	N_2O(g)	219.9
CO(g)	197.7	NO_2(g)	240.1
CO_2(g)	213.7	N_2O_4(g)	304.3
H(g)	114.7	O(g)	161.1
H_2(g)	130.7	O_2(g)	205.1

15. Calculate the standard entropy change of the formation of N_2O_4 using Table 20-1.
    a) -201.7 J/K   b)+322.4 J/K   c) -175.9 J/K
    d) +201.7 J/K   e)+175.9 J/K

16. Calculate the standard entropy change of the formation of ethanol (CH_3CH_2OH) using Table 20-1.
    a) -258.4 J/K   b) -338.7 J/K   c) -171.5 J/K
    d) -346.9 J/K   e) -345.4 J/K

17. Calculate the standard molar entropy of urea ($CO(NH_2)_2$(s)) if the standard entropy change for the formation is -456.3 J/K, use Table 20-1.
    a) -1017.2 J/K·mol   b) +314.1 J/K·mol   c) +194.2 J/K·mol
    d) +104.6 J/K·mol   e) -56.0 J/K·mol

18. The entropy of a substance ___?___ increases as it changes from a liquid to a gas.
    a) never   b) sometimes   c) always
    d) often   e) rarely

19. Entropy generally __?__ with __?__ molecular structure.
    a) increases, increasing   b) decreases, increasing   c) decreases, decreasing
    d) increases, decreasing   e) dissipates, decreasing

20. When a pure liquid or solid dissolves in a solvent, the entropy of the substance generally __?__.

a) decreases   b) increases   c) dissipates

d) changes   e) disappears

21. Which of the following is true about the second law of thermodynamics?

a) $\Delta G = (E^\circ F)\log K$

b) $\Delta S_{uni} + \Delta S_{system} = \Delta S_{surr}$

c) $\Delta G = \Delta H - T\Delta S$

d) $\Delta S_{system} + \Delta S_{surr} = \Delta S_{uni}$

e) $\Delta H = \Delta U + \Delta S$

22. At 298 K, what is the change in entropy of the universe for the system: $A + B \rightarrow AB$ if $\Delta S_{sys} = -108$ J/K and $\Delta H_{sys} = 1500$ kJ? Is the reaction spontaneous?

a) +106.3 J/K, spontaneous

b) -106.3 J/K, spontaneous

c) +1,570 J/K, not spontaneous

d) +1,570 J/K, spontaneous

e) -1,570 J/K not spontaneous

23. The formation $1/2\, A_2 + 2B_2 + C \rightarrow CAB_4$ has an enthalpy of formation of -104 kJ and a change in entropy of -60.8 J/K at 30°C. What is ΔG and spontaneity of the reaction?

a) -85.6 kJ, spontaneous

b) -18.3 kJ, not spontaneous

c) +18.3 kJ, spontaneous

d) +85.6 kJ, not spontaneous

e) -85.6 kJ, not spontaneous

Table 20-2. Enthalpy and Entropy Values.

Compound	ΔS_f°(J/K·mol)	ΔH_f°(kJ/mol)
C(graph)	6	0
Ca(s)	41	0
$CaCO_3$(s)	93	-1207
CO(g)	198	-111
CO_2(g)	214	-394
H_2(g)	131	0
Mg(s)	33	0
MgO(s)	27	-602
N_2(g)	192	0
NH_3(g)	193	-46
O_2(g)	205	0

24. For the formation reaction: $1/2\ N_2$(g) + $3/2\ H_2$(g) $\rightarrow$ NH_3(g) use values of ΔH°_{rxn} and ΔS°_{rxn} to calculate the free energy change at 25 °C. (See Table 20-2).
    a) -53.3 kJ   b) + 29.6 kJ   c) +53.3 kJ
    d) -29.6 kJ   e) -16.5 kJ

25. Using Table 20-2, calculate the free energy change for the formation of carbon dioxide $(CO_2$(g)) using values of ΔH°_{rxn} and ΔS°_{rxn}.
    a) -1,250 kJ   b) -394.4 kJ   c) +458.8 kJ
    d) +853.0 kJ   e) +300 kJ

26. Calculate the free energy change for the combustion of methanol(CH_3OH) to give water and carbon dioxide as products using Table 20-3.
    a) +1,035 kJ   b) -465.2 kJ   c) -702.3 kJ
    d) +536.3 kJ   e) +332.0 kJ

Table 20-3. Free Energy Values.

Compound	ΔG_f°(kJ/mol)
CaO(s)	-604
Ca(OH)$_2$(s)	-899
CH$_3$OH(l)	-166
CH$_3$OH(g)	-163
CO$_2$(g)	-394
HCl(g)	-95
H$_2$O(g)	-229
H$_2$O(l)	-237
NH$_3$(g)	-17
NH$_4$Cl(s)	-203

27. Calculate the free energy for the reaction of hydrochloric acid with ammonia using
Table 20-3.

$$HCl(g) + NH_3(g) \rightarrow NH_4Cl(s)$$

a) -284.9 kJ   b) +100.1 kJ   c) -89.1 kJ
d) +284.9 kJ   e) -100.1 kJ

28. Calculate ΔH, ΔS, and ΔG for the formation of calcium carbonate using Table 20-2 at
298 K. (ΔH, ΔS, and ΔG are given below, respectively)
a) 3600 kJ, -159.3 J/K, 3520 kJ
b) -1207 kJ, -159.3 J/K, -1159 kJ
c) -1207 kJ, 261.9J/K, 76.8 kJ
d) 3600 kJ, 261.9J/K, 3520 kJ
e) -1207 kJ, -261.9 J/K, -1129 kJ

29. If ΔG is negative at all temperatures then ΔS is __?__ and ΔH is __?__ .
a) positive, negative   b) zero, large   c) negative, positive
d) large, zero   e) small, large

30. If ΔG is positive at all temperatures then ΔS is __?__ and ΔH is __?__.

    a) positive, negative   b) negative, positive   c) small, zero

    d) large, zero   e) large, small

31. If ΔH and ΔS are both negative or positive then ΔG has a __?__ sign.

    a) positive   b) negative   c) variable   d) large   e) no

32. At what temperature would a given reaction become spontaneous is $\Delta H = +119$ kJ and $\Delta S = +263$ J/K?

    a) 452 K   b) 2210 K   c) 382 K   d) 2.21 K   e) 363 K

33. Use Table 20-2 to predict when the following reaction becomes spontaneous.

$$MgO(s) + C(graph) \rightarrow Mg(s) + CO(g)$$

    a) 526 K b) 2.48 K   c) 402.5 K   d) 1242 K   e) 2480 K

34. The free energy change for a given reaction is -36.2 kJ. What is the equilibrium constant at 298 K?

    a) 0.985   b) 2.22×10^6 c) 1.01

    d) 8.32×10^{-7} e) 3.25×10^6

35. Use Table 20-3 to calculate the equilibrium constant for the reaction of lime with water at 298 K.

$$CaO(s) + H_2O(l) \rightarrow Ca(OH)_2(s)$$

    a) 3.30×10^{30} b)1.07   c) 3.03×10^{-31}

    d) 1.51×10^6 e)2.01×10^{10}

36. The formation constant for the formation of the complex ion $(AgF_6)^{3-}$ is 5.0×10^{23}. What is the standard free energy change for the process?

    a) +135 kJ   b) -58.7 kJ   c) -135 kJ   d) +58.7 kJ   e)-1956J

37. At what temperature does the equilibrium constant for the formation of methane(CH_4) equal one, when $\Delta H = -74.81$ kJ, $\Delta S = -80.8$ J/K, and $\Delta G = -50.7$ kJ.

    a) 926 K b) 1.08 K   c) 1080 K   d) 762 K   e) 474 K

38. The first law of thermodynamics states that
    a) the energy of every pure substance is zero
    b) disorder is always increasing
    c) enthalpy is always increasing
    d) the total energy of the universe is constant
    e) the entropy of the surroundings is equal to zero.

39. The second law of thermodynamics states that
    a) heat is energy
    b) the enthalpy of the universe is increasing
    c) ΔS of the universe is equal to zero
    d) if ΔG is negative, the reaction is spontaneous
    e) the total entropy of the universe is increasing

40. Which of the following would you expect to have the largest entropy.
    a) $H_2O(l)$ b)$H_2O(g)$ c) $CO(g)$ d) $CO_2(g)$ e)$NH_3(l)$

41. Which of the following would you expect to have the largest entropy.
    a) $He(g)$ b) $CaCO_3(s)$ c)$C_2H_6(g)$ d) $HCl(g)$ e) $HNO_3(l)$

42. Calculate the free energy for the following reaction with the given information.

	$CaCO_3(s)$	$\rightarrow$	$CaO(s)$	+	$CO_2(g)$
ΔS(J/K·mol)	92.9		39.8		213.7
ΔH(kJ/mol)	-1207		-635		-393.5

    a) -47.7 kJ   b) -53.0 kJ   c) +47.7 kJ   d) +53.0 kJ   e) -107 kJ

43. Silver chloride is poorly soluble in water. The solubility constant, Ksp, is 1.8×10^{-10} what is ΔG at 25 °C?
    a) -24.1 kJ   b) -55.6 kJ   c) + 24.1 kJ   d) +55.6 kJ   e) +55.6x10^3 kJ

44. Which of the following statements summarizes the third law of thermodynamics?

   a) The entropy of every pure, perfectly crystalline substance at absolute zero is zero.

   b) The energy of the universe is constant.

   c) The entropy is always increasing.

   d) When ΔH is negative, the reaction is spontaneous.

   e) At higher temperatures a reaction will always occur.

45. The dissociation constant for benzoic acid is 6.3×10^{-5}, what is the free energy for the dissociation?

   a) +30 kJ   b) +10.4 kJ   c) -10.4 kJ

   d) +24x103 kJ   e) +24.0 kJ

46. Arrange $CO_2(g)$, $H_2O(l)$, $Hg(l)$, abd $C(graphite)$ in order of <u>increasing</u> entropy.

   a) $CO_2(g)$, $H_2O(l)$, $Hg(l)$, $C(graphite)$

   b) $C(graphite)$, $Hg(l)$, $H_2O(l)$, $CO_2(g)$

   c) $Hg(l)$, $H_2O(l)$, $C(graphite)$, $CO_2(g)$

   d) $H_2O(l)$, $C(graphite)$, $Hg(l)$, $CO_2(g)$

   e) $Hg(l)$, $CO_2(g)$, $C(graphite)$, $H_2O(l)$

47. Rank $FeO_2(s)$, $CH_4(g)$, $H_2(g)$, and $CH_3OH(l)$ in order of <u>increasing</u> entropy.

   a) $H_2(g)$, $CH_4(g)$, $FeO_2(s)$, $CH_3OH(l)$

   b) $CH_4(g)$, $CH_3OH(l)$, $H_2(g)$, $FeO_2(s)$

   c) $FeO_2(s)$, $CH_3OH(l)$, $H_2(g)$, $CH_4(g)$

   d) $H_2(g)$, $FeO_2(s)$, $CH_3OH(l)$, $CH_4(g)$

   e) $FeO_2(s)$, $H_2(g)$, $CH_3OH(l)$, $CH_4(g)$

48. Which of the following processes would you expect to have a $\Delta S < 0$?

   a) $C(s) + O_2(g) \rightarrow 2 CO(g)$

   b) $CH_3C(CO)H(g) + 5/2 O_2(g) \rightarrow 2 CO_2(g) + 2 H_2O(g)$

   c) $2 NO(g) + O_2(g) \rightarrow NO_2(g)$

   d) $H_2O(g) \rightarrow H_2O(l)$

   e) $CH_3OH(l) \rightarrow CH_3OH(g)$

49. Which of the following shows the greatest increase in disorder?
    a) $NH_4Br(s) \rightarrow NH_3(g) + HBr(g)$
    b) $C_2H_4(g) + HBr(g) \rightarrow C_2H_5Br(g)$
    c) $CO_2(s) \rightarrow CO_2(g)$
    d) $C(s) + 1/2O_2(g) \rightarrow CO(g)$
    e) $C(graphite) + 2H_2(g) \rightarrow CH_4(g)$

50. If a process is exothermic and not spontaneous then what must be true?
    a) $\Delta S > 0$ b) $\Delta H > 0$ c) $\Delta G = 0$ d) $\Delta S < 0$ e) $\Delta H = 0$

51. If ΔS and ΔH are both positive for a given reaction and the reaction is not spontaneous at room temperature, which of the following must be true?
    a) The reaction is spontaneous at high temperatures.
    b) The reaction will never be spontaneous.
    c) ΔH is the driving force.
    d) The reaction is not spontaneous at higher temperatures.
    e) Spontaneity does not depend on temperature.

52. For any reaction at equilibrium, which of the following are true?
    a) $\Delta H < 0$ b) $\Delta S = 0$ c) $\Delta S < 0$ d) $\Delta H = 0$ e) $\Delta G = 0$

53. Which of the following is true about vaporization?
    a) ΔS is positive and ΔH is negative
    b) ΔS, ΔH and ΔG are all negative
    c) ΔS and ΔH are both negative
    d) ΔS and ΔH are both positive at all temperatures
    e) ΔS, ΔH, and ΔG are equal to zero

54. If a forward reaction has $\Delta G > 0$, which of the following are true?
    a) ΔS and ΔH are both positive
    b) ΔS and ΔH are negative
    c) ΔS is positive and ΔH is negative
    d) the reaction will not occur under any conditions
    e) the reverse reaction is spontaneous

55 . All of the following have $\Delta G = 0$ EXCEPT

   a) $O_2(g)$ b) $Br_2(g)$ c) $H_2(g)$ d) $Ca(s)$ e) $Hg(l)$

56. Which of the following does not have a free energy of zero?

   a) $N(g)$ b) $I_2(s)$ c) $Fe(s)$ d) $He(g)$ e) $Ne(g)$

Chapter 20 : Answers to Multiple Choice

11. a	21. d	31. c
12. c	22. d	32. a
13. e	23. a	33. e
14. b	24. e	34. b
15. a	25. b	35. a
16. e	26. c	36. c
17. d	27. a	37. a
18. c	28. e	38. d
19. a	29. a	39. e
20. b	30. b	40. b

41. c	51. a
42. a	52. e
43. d	53. d
44. a	54. e
45. e	55. b
46. b	56. a
47. c	57.
48. c	58.
49. a	59.
50. d	60.

Chapter 21
Electron Transfer Reactions

Section A: Free Response

Use the following electrochemical cell figure with the reaction that takes place to answer question 1.

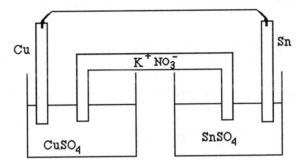

Figure 1. Electrochemical Cell

$$Cu^{2+} + Sn(s) \rightarrow Cu(s) + Sn^{2+}$$

1. What are the two half reactions occurring in the cell? Which reaction occurs at the anode? the cathode? Describe how the electrons flow and the role the salt bridge plays in the cell.

2. A Downs cell is used to produce Na by the electrolysis of molten NaCl. What is added to the molten NaCl and why? What is the other product.

3. Explain the corrosion of iron and write out the reaction that occurs when iron is exposed to oxygen and water.

4. Write the net reaction of a lead storage battery. Lead batteries are used in automobiles, yet they last for months at a time. Explain how car batteries can last, without going dead.

5. Identify the element oxidized and element reduced in the following reaction which occurs in acidic solution. Balance the equation.

$$I^-(aq) + MnO_4^-(aq) \rightarrow IO_3^-(aq) + MnO_2(s)$$

6. Identify the element oxidized and element reduced in the following reaction which occurs in acidic solution. Balance the equation.

$$I^-(aq) + MnO_4^-(aq) \rightarrow IO_3^-(aq) + MnO_2(s)$$

7. Balance the following redox equation which occurs under acidic conditions in aqueous solution:

$$As_2O_3(s) + MnO_4^-(aq) \rightarrow H_3AsO_4(aq) + Mn^{2+}(aq)$$

8. Balance the following redox equation which occurs under basic conditions in aqueous solution.

$$Se \rightarrow SeO_3^{2-} + Se^{2-}$$

9. Classify each of the following reactions as "redox" or "not redox." In each case of redox identify the substance oxidized and the substance reduced. Write "not redox"
   a) $2\,HgO(s) \rightarrow 2\,Hg(l) + O_2(g)$
   b) $H_2O(l) + SO_3(g) \rightarrow H_2SO_4(aq)$
   c) $2\,HCl(aq) + Mg(s) \rightarrow H_2(g) + MgCl_2(aq)$
   d) $H_2CO_3(aq) \rightarrow H_2O(l) + CO_2(g)$
   e) $Ba(NO_3)_2(aq) + Na_2SO_4(aq) \rightarrow BaSO_4(s) + 2\,NaNO_3(aq)$

10. A student prepared an acidic solution of $K_2Cr_2O_7$ and noted its color as yellow. Gaseous SO_2 was bubbled into the yellow solution. Immediately the color of the solution changed to aqua blue indicating the presence of Cr^{3+}. A student was asked to write the balanced equation for this reaction. Uncertain of the sulfur containing product, the student removed a small amount of the mixture, and treated it with a few drops of barium chloride. A white precipitate formed. The precipitate was identical to that formed when barium chloride is added to sulfuric acid. Now the student was able to write the balanced net-ionic equation, identify the substance oxidized and the substance reduced. You are asked to do the same and discuss the logic of the experiment.

Key Concepts for Free Response:

1. $Sn(s) \rightarrow Sn^{2+} + 2e^-, Cu^{2+} + 2e^- \rightarrow Cu(s)$
   are the two half reactions occurring in the cell. Copper is reduced at the cathode and tin is oxidized at the anode. The electrons flow toward the cathode, copper. The salt bridge serves as a ion transporter, the copper ion is taken out of solution and the counter ion travels through the salt bridge to the tin cell. The tin is being oxidized and is going into the solution, therefore it has to have a counter ion.

2. $BaCl_2$ and $CaCl_2$ are added to the dry NaCl to lower the melting point from 800°C to 600°C, therefore not as much energy is required. Chlorine gas is also produced.

3. When iron corrodes, the metal is converted into rust, which consists of hydrated iron(III)oxide.
   $$2Fe(s) + 2H_2O(l) + O_2(g) \rightarrow 2Fe(OH)_2(s)$$
   If oxygen is not freely available , the iron(II)hydroxide is further oxidized to form iron oxide. This reaction is about 100 times faster than the corrosion that takes place without oxygen present. There has to be an anodic and a cathodic area for corrosion to occur. The anode is generally at cracks in the oxide coating or around impurities. The cathodic are as occur where there is metal oxide coating or at impurity sites. When there is an electrical connection between the two and an electrolyte, such as hydroxide ion, that both are in contact, corrosion can and will occur.

4. $Pb(s) + PbO_2(s) + 2H_2SO_4(aq) \rightarrow 2PbSO_4(s) + 2H_2O(l)$
   Electrical energy can be added to the battery to reverse the net process, thus recharging the battery. The lead(II)sulfate is converted back to lead and lead(IV)oxide and sulfuric acid is produced. This is in effect electrolysis.

5. Iodine is oxidized and manganese is reduced.
   $$2 H^+(aq) + I^-(aq) + 2 MnO_4^-(aq) \rightarrow IO_3^-(aq) + 2 MnO_2(s) + H_2O$$

6. Iodine is oxidized and manganese is reduced.
   $$2 H^+(aq) + I^-(aq) + 2 MnO_4^-(aq) \rightarrow IO_3^-(aq) + 2 MnO_2(s) + H_2O$$

7. $9 H_2O + 12 H^+ + 5 As_2O_3 + 4 MnO_4^- \rightarrow 10 H_3AsO_4 + 4 Mn^{2+}$

8. $6 OH^- + 3 Se \rightarrow SeO_3^{2-} + 2 Se^{2-} + 3 H_2O$

9. a) redox. Hg^{2+} is reduced to Hg. O^{2-} is oxidized to O_2. b) not redox c) redox. H^+ is reduced to H_2. Mg is oxidized to Mg^{2+} d) not redox e) not redox

10. When the yellow solution changes to aqua blue, the $Cr_2O_7^{2-}$ is reduced to Cr^{3+}. Simultaneously the SO_2 is oxidized to SO_4^{2-}. This oxidation is verified by the precipitation of BaSO4 when a few drops of barium chloride is added to the reaction mixture. The balanced reaction is
    $$4 H^+ + 3 SO_2 + 2 CrO_4^{2-} \rightarrow 3 SO_4^{2-} + 2 Cr^{3+} + 2 H_2O$$

Section B: Multiple Choice

11.   Which of the following is the correct cell notation for the reaction
$$Hg_2^{2+} \ + \ Cd(s) \ \rightarrow \ Cd^{2+} \ + \ 2\,Hg(l)$$
a) $Cd^{2+}\,|\,Cd \ \| \ Hg_2^{2+}\,|\,Hg$
b) $Cd^{2+}\,|\,Hg_2^{2+} \ \|\,Cd \ |\,Hg$
c) $Cd\,|\,Cd^{2+} \ \| \ Hg_2^{2+}\,|\,Hg$
d) $Cd^{2+}\,|\,Hg \ \| \ Hg_2^{2+}\,|\,Cd$
e) $Hg\,|\,Cd \ \| \ Hg_2^{2+}\,|\,Cd^{2+}$

12.   Which of the following is the correct cell notation for the reaction
$$Au^{3+} \ + \ Al(s) \ \rightarrow \ Al^{3+} \ + \ Au(s)$$
a) $Al^{3+}\,|\,Al \ \| \ Au^{3+}\,|\,Au$
b) $Al \ |\,Al^{3+} \ \| \ Au^{3+}\,|\,Au$
c) $Al^{3+}\,|\,Au^{3+} \ \|\,Al \ |\,Au$
d) $Al^{3+}\,|\,Au \ \| \ Au^{3+}\,|\,Al$
e) $Au\,|\,Al \ \| \ Au^{3+}\,|\,Al^{3+}$

Table 21-1: Standard Reduction Potentials in Aqueous Solution at 25°C

Acidic Solution	E° (volts)
$Au^{3+}(aq) \ + \ 3\,e^- \rightarrow \quad Au(s)$	+1.50
$Br_2(l) \quad + \ 2\,e^- \rightarrow \quad 2\,Br^-(aq)$	+1.80
$Ag^+\,(aq) \ + \ e^- \rightarrow \quad Ag(s)$	+0.80
$I_2(s) \quad\quad + \ 2\,e^- \rightarrow \quad 2\,I^-\,(aq)$	+0.535
$Sn^{2+}(aq) \quad + \ 2\,e^- \rightarrow \quad Sn(s)$	–0.14
$Zn^{2+}(aq) \quad + \ 2\,e^- \rightarrow \quad Zn(s)$	–0.763

13. Using the information in Table 21-1 calculate ΔG for the following reaction:
$$2\,Ag^+\,(aq) \ + Sn(s) \ \rightarrow \quad Cu(s) + Sn^{2+}\,(aq)$$
a) +64 kJ b) +91 kJ   c) +181 kJ   d) -64 kJ   e) –181 kJ

14. Using the information in Table 21-1 calculate ΔG for the following reaction:

$$I_2(s) \;+\; 2\,Br^-(aq) \;\rightarrow\; 2\,I^-(aq) \;+\; Br_2(l)$$

a) +105 kJ   b) -105 kJ   c) +312 kJ   d) +52 kJ   e) –312 kJ

15. Using the information in Table 21-1 calculate ΔG for the following reaction?

$$2\,Au^{3+}(aq) \;+\; 3\,Zn(s) \;\rightarrow\; 2\,Au(s) \;+\; 3\,Zn^{2+}(aq)$$

a) +1320 kJ   b) +660 kJ   c) -428 kJ   d) –1320 kJ   e) -660 kJ

16. If ΔG of following reaction is -203 kJ, what is E^o?

$$2Ag^+(aq) + Ni(s) \;\rightarrow\; 2Ag(s) + Ni^{2+}(aq)$$

a) -1.05v b) +2.10v   c) +0.0011v   d) -0.0011v   e) +1.05v

17. If ΔG of the following reaction is -114 kJ what is E^o?

$$A^{3+}(aq) + 3B(s) \;\rightarrow\; A(s) + 3B^+(aq)$$

a) +0.59v   b) -0.09v   c) -0.59v   d) +0.00059v e) +0.09v

18. Given the two half reactions and their potentials, which net reaction is spontaneous?

$$Mg^{2+}(aq) + 2e^- \;\rightarrow\; Mg(s) \qquad E^o = -2.37v$$
$$Ni^{2+}(aq) \;+\; 2e^- \;\rightarrow\; Ni(s) \qquad E^o = -0.25v$$

a) $Ni(s) + Mg^{2+}(aq) \;\rightarrow\; Mg(s) + Ni^{2+}(aq)$
b) $Ni^{2+}(aq) + Mg(s) \;\rightarrow\; Mg^{2+}(aq) + Ni(s)$
c) $Ni(s) + Mg(s) \;\rightarrow\; Mg^{2+}(aq) + Ni^{2+}(aq)$
d) $Mg^{2+}(aq) + Ni^{2+}(aq) \;\rightarrow\; Mg(s) + Ni(s)$
e) $Mg^{2+}(aq) + Mg(s) \;\rightarrow\; Ni(s) + Ni^{2+}(aq)$

19. Given the two half reactions and their potentials, which net reactions not spontaneous?

$$Zn^{2+}(aq) + 2e- \rightarrow Zn(s)$$
$$Ni^{2+}(aq) + 2e- \rightarrow Ni(s)$$

a) $Ni^{2+}(aq) + Zn^{2+}(aq) \rightarrow Ni(s) + Zn(s)$

b) $Ni^{2+}(aq) + Zn(s) \rightarrow Ni(s) + Zn^{2+}(aq)$

c) $Ni(s) + Zn^{2+}(aq) \rightarrow Zn(s) + Ni^{2+}(aq)$

d) $Zn(s) + Ni(s) \rightarrow Ni^{2+}(aq) + Zn^{2+}(aq)$

e) $Zn(s) + Zn^{2+}(aq) \rightarrow Ni^{2+}(aq) + Ni(s)$

20. Using Table 21-2, calculate E^o for the following reaction.

$$Sn^{4+}(aq) + 2K(s) \rightarrow Sn^{2+}(aq) + 2 K^+(aq)$$

a) +6.00v   b) -3.075v   c) +3.075v   d) +2.775v   e) -2.775v

Table 21-2: Standard Reduction Potentials in Aqueous Solution at 25°C

Acidic Solution			E^o (volts)
$F_2(s)$	$+ 2e^- \rightarrow$	$2F^-$ (aq)	+2.87
$Cl_2(g)$	$+ 2e^- \rightarrow$	$2Cl^-(aq)$	+1.36
$Sn^{4+}(aq)$	$+ 2e^- \rightarrow$	$Sn^{2+}(s)$	+0.15
$Cd^{2+}(aq)$	$+ 2e^- \rightarrow$	$Cd(s)$	–0.40
$Al^{3+}(aq) + 3e^- \rightarrow$		$Al(s)$	-1.66
K^+ (aq) $+ e^- \rightarrow$		$K(s)$	-2.93

21. Using Table 21-2, calculate E^o for the following reaction.

$$2F^-(aq) + Cl_2(g) \rightarrow F_2(g) + 2Cl^-(aq)$$

a) -1.51v   b) +8.46v   c) -4.23v   d) -8.46v   e)+4.23v

22. Using Table 21-2, calculate E^o for the following reaction.

$$2Al^{3+}(aq) + 3Cd(s) \rightarrow 2Al(s) + 3Cd^{2+}(aq)$$

a) -2.06v   b) +4.52v   c) +2.06v   d) -4.52v   e) -1.26v

23. Use Table 21-2 and the reaction

$$2Na(s) + F_2(g) \rightarrow 2F^-(aq) + 2Na^+(aq) \qquad E^0 = +5.58v$$

to calculate the standard oxidation potential of the half reaction

$$Na(s) \rightarrow Na^+(aq) + e^-$$

a) -1.36v   b)+8.45v   c) -2.71v   d) +2.71v   e) -8.45v

24. Use Table 21-1 and the reaction

$$Pt(s) + 2Ag^+(aq) \rightarrow Pt^{2+}(aq) + 2Ag(s)$$

to calculate the standard reduction potential of the half reaction

$$Pt^{2+}(aq) + 2e^- \rightarrow Pt(s)$$

a) +1.20v   b) +0.40v   c) -1.20v   d) +2.00v   e) -2.00v

25. If an electrochemical cell with the notation Rb|Rb$^+$||Na$^+$|Na, which has a standard potential of +0.21v, what is the standard reduction potential of the Rb half cell is the Na$^+$/Na half reaction has a reduction potential of -2.71v?

a) +2.50v   b) -2.50v   c) -2.94v   d) +2.94v   e) 2.30v

26. What is the equilibrium constant for the following reaction at 298 K?

$$2Ag^+(aq) + 2I^-(aq) \rightarrow I_2(s) + 2Ag(s) \qquad E^0 = +0.265v$$

a) 2.99×10^4 b) 8.97×10^8 c) 7.73×10^3 d) 87.9   e) 1.60×10^7

27. What is the equilibrium constant for the following reaction at 37°C?

$$Hg_2^{2+}(aq) + 2Cl^-(aq) \rightarrow 2Hg(l) + Cl_2(g)$$

a) 5.12×10^{-20} b) 1.74×10^{-43} c) 2.12×10^{28} d) 2.69×10^{-19} e) 1.95×10^{19}

28. What is the equilibrium constant for the following reaction at 20 °C?

$$Fe(s) + Cu^{2+}(aq) \rightarrow Fe^{2+}(aq) + Cu(s) \qquad E^0 = +0.78v$$

a) 8.59×10^{20} b) 2.25×10^{26} c) 1.45×10^{27} d) 1.02   e) 1.16×10^{-21}

29. What is the cell potential for

$$Fe(s) + Cd^{2+}(aq) \rightarrow Fe^{2+}(aq) + Cd(s) \qquad E^0 = 0.040v$$

when [Fe^{2+}] = 0.020 and [Cd^{2+}] = 0.20 at 298 K?

a) +0.019v   b) +0.039v   c) +0.010v   d) +0.099v   e) +0.069v

30. What is the cell potential for

$$3Sn^{4+}(aq) + 2Al(s) \rightarrow 3Sn^{2+}(aq) + 2Al^{3+}(aq)$$
when $[Sn^{4+}] = 1.0$, $[Sn^{2+}] = 2.0$, and $[Al^{3+}] = 1.5$ at 298 K.
a) -1.82v   b) +1.82v   c) +1.81v   d) +1.80v   e) +1.60v

31. Calculate the cell potential for the following reaction at 30 °C.
$$Zn(s) + 2Co^{3+}(aq) \rightarrow Zn^{2+}(aq) + 2Co^{2+}(aq) \ E^\circ = 2.58v$$
if $[Zn^{2+}] = 0.10$, $[Co^{3+}] = 0.05$, and $[Co^{2+}] = 0.50$.
a) +2.56v   b) +2.55v   c)+2.61v   d)+2.58v   e)-2.61v

32. The equilibrium constant of the cell $Pb^{2+}|Pb^{4+}||Pt^{2+}|Pt$ at 25 °C is 5.37×10^{-21}, what is the cell potential?
a)-0.60   b) -1.38   c) -1.20   d) +1.38   e) +0.60

33. The equilibrium constant for the following reaction is 1.65×10^{16} at 298 K, what is E°?
$$MnO_2(s) + 4H^+(aq) + 2Cl^-(aq) + SbCl_4^-(aq) \rightarrow$$
$$Mn^{2+}(aq) + SbCl_6^-(aq) + 2H_2O(l)$$
a) +1.11v   b) + 0.96v   c)+0.48v   d) -0.86v   e) -0.43v

34. If the potential of a cell is +1.32v at Q = 0.0969 with n = 2, what is the standard potential of the cell?
a) +1.35v   b) +1.48v   c) +1.31v   d) +1.34v   e) +1.29v

35. The potential of the following cell is +3.12v when $[Cu^{2+}] = 0.10$ and $[Na^+] = 0.02$, what is E°?
$$Cu^{2+}(aq) + 2Na(s) \rightarrow Cu(s) + 2Na^+(aq)$$
a) +3.05v   b) +3.10   c) +3.19v   d) +3.14v   e) +2.96v

Table 21-3: Standard Reduction Potentials in Aqueous Solution at 25°C

Acidic Solution	$E°$ (volts)
$Cl_2(g)$ + 2 e^- → 2 Cl^-(aq)	+1.36
$Br_2(s)$ + 2 e^- → 2Br^- (aq)	+1.08
$I_2(s)$ + 2 e^- → 2I^- (aq)	+0.535
Cu^{2+}(aq) + 2 e^- → Cu(s)	+0.337
$2H_2O(l) + 2e^-$ → $H_2(g) + 2OH^-$(aq)	-0.828
K^+ (aq) + e^- → K(s)	-2.93
Li^+(aq) + 1 e^- → Li(s)	-3.045

36. Use Table 21-3 to predict the products at the cathode when electric current is passed through a solution of KI.

a) K(s)   b) $H_2(g)$ c) $I_2(l)$ d) $O_2(g)$ e) $H_2O(l)$

37. Use Table 21-3 to predict the products at the anode of the same situation as the previous question.

a) $I_2(l)$ b) K^+(aq)   c) $H_2(g)$ d) K(s)   e) $O_2(g)$

38. If electric current is passed through molten $CuCl_2$, the product at the anode would be
____?____ and the product at the cathode would be ____?____ (use Table 21-3).

a) $H_2O(l)$, $Cl_2(g)$ b) Cu(s), $Cl_2(g)$ c) $H_2(g)$, Cu(s)

d)$Cl_2(g)$, Cu(s)   e) Cu(s), $H_2O(l)$

39. If electric current is passed though aqueous LiBr, the product at the cathode would be
____?____ and the product at the anode would be ____?____ (use Table 21-3).

a) $H_2O(l)$, Li^+(aq)   b) $Br_2(l)$, Li(s)   c) Li(s), $Br_2(l)$

d) $Br_2(l)$, $H_2(g)$ e) $H_2(g)$, $Br_2(l)$

40. How long would it take to deposit 1.36g of copper from an aqueous solution of copper(II)sulfate by passing a current of 2 amperes through the solution?

a) 2066 sec   b) $1.11x10^{-5}$ sec   c) 2567 sec

d) 736 sec   e) 1033 sec

41. If you wish to plate 1.5 g of gold by electrolyzing a solution of Au^{3+} with 2.5 amperes, how long would it take?

a) 57,900 sec   b) 882 sec   c) 2.09×10^{-8} sec

d) 500 sec   e)294 sec

42. If a current of 6 amps is passed through a solution of Ag^+ for 1.5 hours, how many grams of silver are produced?

a) 0.604g   b) 36.2g   c) 0.335g   d) 3.04g   e) 1.00g

43. How much Platinum would be produced by passing a 2 ampere current through a solution of Pt^{2+} for 30 minutes?

a)7.27g   b)0.121g   c)1.82g   d) 2.82g   e) 74.5g

44. How many kilowatt hours of electrical energy are required to plate 2 grams of silver form an aqueous solution of silver nitrate onto a necklace using 3.00 v.

a) 0.00147 kwh   b) 0.000165 kwh   c) 32.4 kwh

d) 0.00149 kwh   e) 2.07 kwh

45. How many kwh of electrical energy are required to produce 1 kg of Aluminum from a solution of Al^{3+} using 2.5 v?

a) 7.45 kwh   b) 2.48 kwh   c) 0.397 kwh

d) 1.19 kwh   e) 0.000382 kwh

46. How was aluminum originally made?

a) The Hall-Heroult process

b) Al_2O_3 mixed with cryolite is electrolyzed

c) electrolysis of molten Al_2O_3

d) mining and purifying directly

e) reducing $AlCl_3$ with sodium

47. What metal is utilized at the anode of a mercury battery?

a)Pb   b) Hg   c) Zn   d) Ni   e)Pt

Use the Tables 21-1 through 21-3 to answer questions 51 and 52.

48. Which of the following is the best oxidizing agent?

 a) F_2 b) Ag c) Ni^{2+} d) Ag^+ e) Al^{3+}

49. Which of the following is the best reducing agent?

 a) Ag^+ b) Al c) F^- d) Ni^{2+} e) F_2

50. Under acidic conditions the bromate ion is reduced to the bromide ion. Write the balanced half reaction for this process.

 a) $BrO_3^- + 6\ H^+ + 6\ e^- \rightarrow Br^- + 3\ H_2O$
 b) $2\ BrO_3^- + 6\ H^+ \rightarrow Br_2^- + 6\ H_2O + 3\ e^-$
 c) $BrO_3^- + 6\ H_2O + 10\ e^- \rightarrow Br_2^- + 12\ H^+ + 3\ O_2$
 d) $2\ BrO_3^- + 6\ H_2O \rightarrow 2\ Br^- + 12\ H^+ + 6\ O_2 + 8\ e^-$
 e) $2\ BrO_3^- + 6\ H^+ \rightarrow Br_2^- + 3\ H_2O + 3\ e^-$

51. Balance the following redox equation which occurs in acidic solution.

$$Cu_{(s)} + NO_3^-{}_{(aq)} \rightarrow Cu^{2+}{}_{(aq)} + NO_{(g)}$$

 a) $4\ H^+ + NO_3^- + Cu(s) \rightarrow Cu^{2+} + NO + 2\ H_2O$
 b) $2\ H_2O + NO_3^- + Cu(s) \rightarrow NO + Cu^{2+} + 4\ H^+$
 c) $2\ NO_3^- + 8H^+ + 3Cu(s) \rightarrow 3\ Cu^{2+} + 2\ NO + 4\ H_2O$
 d) $3\ NO_3^- + 6H^+ + 2Cu(s) \rightarrow 2\ Cu^{2+} + NO + 3\ H_2O$
 e) $6\ H^+ + 3\ NO_3^- + 2Cu^{2+} \rightarrow Cu(s) + NO + 3\ H_2O$

52. Balance the following redox equation which occurs in acidic solution.

$$N_2H_{4(g)} + BrO_3^-{}_{(aq)} \rightarrow Br^-{}_{(aq)} + N_{2(g)}$$

 a) $3\ N_2H_4 + BrO_3^- \rightarrow 3\ N_2 + Br^- + 3\ H_2O + 6\ H^+$
 b) $N_2H_4 + BrO_3^- + 2\ H^+ \rightarrow 2\ Br^- + N_2 + 3\ H_2O$
 c) $3\ N_2H_4 + 2\ BrO_3^- + 12\ H^+ \rightarrow 3\ N_2 + 2\ Br^- + 6\ H_2O + 12\ H^+$
 d) $N_2H_4 + 2\ BrO_3^- + 8\ H^+ \rightarrow 2\ Br^- + N_2 + 6\ H_2O +$
 e) $3\ N_2H_4 + 2\ BrO_3^- \rightarrow 3\ N_2 + 2\ Br^- + 6\ H_2O$

53. What is the balanced reduction half reaction under acidic conditions in the equation below?

$$SO_2 + Cr_2O_7^{2-} \longrightarrow Cr^{3+} + SO_4^{2-}$$

a) $14\ H^+ + Cr_2O_7^{2-} \longrightarrow 2\ Cr^{3+} + 7\ H_2O + 8e^-$

b) $14\ H^+ + Cr_2O_7^{2-} \longrightarrow 2\ Cr^{3+} + 7\ H_2O + 6e^-$

c) $8e^- + 14\ H^+ + Cr_2O_7^{2-} \longrightarrow 2\ Cr^{3+} + 7\ H_2O$

d) $6e^- + 14\ H^+ + Cr_2O_7^{2-} \longrightarrow 2\ Cr^{3+} + 7\ H_2O$

e) $9e^- + 14\ H^+ + Cr_2O_7^{2-} \longrightarrow Cr^{3+} + 7\ H_2O$

54. Which of the following reactions is NOT a redox reaction?

a) $2\ HgO(s) \longrightarrow 2\ Hg(l) + O_2\ (g)$

b) $H_2\ (g) + Br_2\ (g) \longrightarrow 2\ HBr(g)$

c) $2\ HCl\ (aq) + Zn\ (s) \longrightarrow H_2\ (g) + ZnCl_2(aq)$

d) $H_2CO_3(aq) \longrightarrow H_2O(l) + CO_2(g)$

e) $2\ KClO_3 \longrightarrow 2\ KCl(s) + 3\ O_2(g)$

55. Balance the following redox reaction which occurs in acidic solution:

$$Sn_{(s)} + NO_3^-{}_{(aq)} \longrightarrow Sn^{4+}{}_{(aq)} + N_2O_{(g)}$$

a) $Sn(s) + 10H^+ + 2NO_3^- \longrightarrow N_2O + Sn^{4+} + 5H_2O$

b) $2Sn(s) + 10H^+ + 2NO_3^- \longrightarrow N_2O + 2Sn^{4+} + 5H_2O$

c) $3Sn(s) + 4H^+ + 2NO_3^- \longrightarrow 2N_2O + 3Sn^{4+} + 2H_2O$

d) $2Sn(s) + 2H^+ + 2NO_3^- \longrightarrow N_2O + 2Sn^{4+} + H_2O$

e) $Sn(s) + 6H^+ + 2NO_3^- \longrightarrow N_2O + 2Sn^{4+} + 3H_2O$

Chapter 21 : Answers to Multiple Choice

11. c	21. a	31. b
12. b	22. e	32. a
13. e	23. d	33. c
14. a	24. a	34. e
15. d	25. c	35. a
16. e	26. b	36. b
17. a	27. d	37. a
18. b	28. a	38. d
19. c	29. e	39. e
20. c	30. d	40. a

41. b	51. c
42. b	52. e
43. e	53. d
44. d	54. d
45. a	55. b
46. e	
47. c	
48. a	
49. b	
50. a	

Chapter 22
The Chemistry of the Main Group Elements

Section A: Free Response

1. Outline a useful laboratory method to prepare the elements H_2, O_2, and Cl_2

2. Compare the physical properties of carbon and silicon oxides with respect to their bonding patterns.

3. Write equations for the following laboratory procedure for the behavior of magnesium metal when burned in air.

> Magnesium metal is burned in air to produce a mixture of two solids, both binary compounds. In order to get a pure product, one of the binary compounds can be converted into the other. The mixture is treated with water. Upon strong heating ammonia gas can be detected. The final product is pure.

4. Explain why F_2 is a better oxidizing agent that Cl_2 in the gas phase.

5. What is a disproportionation reaction? Discuss the disproportionation reaction which occurs when chlorine gas is dissolved in water. Explain how the reaction is pH dependent.

Key Concepts for Multiple Choice

1. a) Among the methods that hydrogen can be prepared is the action of a metal with an acid such as

    $$2 \, HCl \, (aq) \; + \; Zn \, (s) \rightarrow \; H_2 \, (g) \; + \; ZnCl_2(aq)$$

    b) Among the methods that oxygen can be prepared is the decomposition of potassium chlorate

    $$KClO_3(s) \; \rightarrow \; 2 \, KCl(s) \; + \; O_2 \, (g)$$

    c) Among the methods that chlorine can be prepared is the electrolysis of NaCl and the oxidation of NaCl under acidic conditions with potassium dichromate.

    $$2 \, NaCl(s) \; \text{--- electrolysis-->} \; 2 \, Na(s) \quad + \; Cl_2 \, (g)$$

2. Carbon dioxide is a gas at room conditions. It exists as molecules with a double bond between each carbon and oxygen. In the case of SiO_2 the empirical formula not represent a molecule, but the ratio of atoms in a network. The high melting point supports this observation. Each silicon atom is singly bonded to four oxygen atoms which in turn bond to other silicon atoms.

3. Air is a mixture of O_2 and N_2 so two reactions may occur

    Reaction 1: $2 \, Mg(s) + O_2(g) \text{---> } 2 \, MgO(s)$

    Reaction 2: $3 \, Mg(s) + N_2(g) \text{---> } Mg_3N_2(s)$

    When the mixture is treated with water the Mg_3N_2 will react according to

    $$Mg_3N_2 \, (s) \quad + 6 \, H_2O(l) \text{---> } 3 \, Mg(OH)_2(s) \; + \; 2 \, NH_3(g)$$

    With heating the $Mg(OH)_2$ will decompose by the reaction

    $$Mg(OH)_2 \text{---> } MgO(s) \; + \; H_2O(g)$$

    Thus the Mg_3N_2 has been totally converted to MgO. The product in the crucible now is pure MgO.

4. The main reason is that the F_2 bond is much weaker than Cl_2. In the gas phase, electronegativities are not a major factor as they are in solution.

5. A disproportionation reaction is a redox reaction in which the same element is oxidized and reduced. In this case

    $$Cl_2(g) \; + \; 2 \, H_2O(l) \; \rightleftharpoons \; H_3O^+(aq) \; + \; HOCl(aq) \; + \; Cl^-(aq)$$

    One chlorine atom is oxidized from 0 to +1 in the hypochlorite ion and the other chlorine atom is reduced to the chloride ion. At low pH the equilibrium is shifted toward the unionized molecules and at higher pH the equilibrium is shifted toward the chloride ion and hypochlorous acid.

Section B: Multiple Choice

6. Of the elements N, P, As, Sb, and Bi, which one has the most metallic character?

    a) N   b) P   c) As   d) Sb   e) Bi

7. A group of metals which all melt below 200°C are

    a) alkali   b) transition   c) alkaline earth   d) precious   e) actinide

8. Which property of metals decreases as one moves down a group in the periodic chart?

    a) atomic radius   b) ionic radius   c) ionization energy

    d) atomic mass   e) atomic number

9. The primary product of the reaction of sodium with pure oxygen is not the anticipated
   sodium oxide. It is ___?___ with the formula ___?___.

    a) sodium peroxide, Na_2O b) sodium peroxide, NaO_2

    c) sodium superoxide, Na_2O_2 d) sodium peroxide, Na_2O_2

    e) sodium superoxide, Na_2O_2

10. Potassium superoxide can be used to produce oxygen under very controlled conditions
    according to the equation:

    a) $2 KO(s) + CO_2(g) \rightarrow K_2CO_3(s) + 1/2 O_2 (g)$

    b) $K_2O_2(s) + CO_2(g) \rightarrow K_2CO_3(s) + 1/2 O_2 (g)$

    c) $4 KO_2(s) + 2 CO_2(g) \rightarrow 2 K_2CO_3(s) + 3 O_2 (g)$

    d) $KO_3(s) + 2 CO_2(g) \rightarrow 2 K_2CO_3(s) + 7/2 O_2 (g)$

    e) $2 KO_4(s) + 2 CO_2(g) \rightarrow 2 K_2CO_3(s) + 3 O_2 (g)$

11. Synthesis gas (syngas) can be produced from coal gas to produce which of the following free
    elements?

    a) oxygen   b) carbon   c) hydrogen   d) helium   e) sulfur

12. Which of the following methods are useful for the production of hydrogen?

    a) metal + acid   b) carbonate + acid   c) acid + base

    d) acid + alcohol   e) all of these

13. The largest use of hydrogen gas is for the industrial production of
   a) aspirin   b) alcohol   c) gasoline   d) sodium   e) ammonia

14. The main compound in blackboard chalk is _____ which is also an ingredient in
   a) $CaCO_3$, stomach remedies   b) $MgCO_3$, egg shells   c) NaCl, table salt
   d) $Mg(OH)_2$, milk of magnesium   e) SiO_2, sand

15. The second most abundant element in the earth's crust is
   a) aluminum   b) hydrogen   c) silicon   d) sulfur   e) carbon

16. Oxides of the alkaline earth family form
   a) basic solutions   b) acidic solutions   c) gases with water
   d) noble gas compounds   e) soluble sulfides

17. The equation for the Haber Process is
   a) $N_2(g) + 3 H_2 (g) \rightleftharpoons 2 NH_3 (g)$
   b) $CH_4 (g) + 2 O_2 (g) \rightarrow CO_2(g) + 2 H_2O(g)$
   c) $4 CS_2 (g) + 4 O_2 (g) \rightleftharpoons 4 CO_2(g) + S_8(s)$
   d) $N_2H_4 (g) + O_2 (g) \rightarrow N_2 (g) + 2 H_2O(g)$
   e) $NH_3(aq) + NaOCl(aq) \rightarrow N_2H_4 (aq) + NaCl(aq) + H_2O (l)$

18. How many oxides of nitrogen are have been characterized by chemists?
   a) two   b) three   c) four   d) five   e) more than six

19. The allowable positive oxidation numbers of nitrogen are
   a) +2, +4   b) +2, +4, +6   c) +1, +3, +5
   d) +2, +4, +5   e) +1, +2, +3, +4, +5

20. Which gases are commercially prepared from the liquefaction of air?
   a) nitrogen   b) oxygen   c) nitrogen and oxygen
   d) helium and nitrogen   e) hydrogen and oxygen

21. The Ostwald process is useful for the preparation of
   a) ammonia from nitrogen and hydrogen   b) sulfur from iron sulfide
   c) nitric acid from ammonia   d) oxygen from sand
   e) lead from lead sulfide

22. The building block of the silicate minerals is the
   a) tetrahedral SiO_4 unit   b) the bent SiO_2 unit   c) the bent SiO_4 unit
   d) the trigonal planar SiO_3 unit   e) the octahedral SiO_4 unit

23. Which of the following metals is NOT attached by nitric acid?
   a) Fe   b) Ti   c) Au   d) Cu   e) Co

24. Aluminum does not react with nitric acid because
   a) aluminum is a noble metal
   b) the surface of aluminum is coated with unreactive Al_2O_3
   c) aluminum will not gain electrons
   d) aluminum forms slightly acidic hydrates
   e) aluminum usually contains a trace amounts of other metals such as Cr^{3+}

25. Which compound ranks #1 in terms of pounds produced annually in the United States?
   a) ethanol   b) ammonia   c) benzene
   d) sulfuric acid   e) sodium hydroxide

Chapter 22 : Answers to Multiple Choice

6. e	16. a
7. a	17. a
8. c	18. e
9. d	19. e
10. c	20. c
11. c	21. c
12. a	22. a
13. e	23. c
14. a	24. b
15. c	25. d

Chapter 23
The Chemistry of the Transition Elements

Section A: Free Response

1. Discuss the trends in densities and melting points of the transition metals.

2. List three methods that are used in the purification of copper metal. Describe one of these methods in more detail.

3. Write chemical formula or the structure for each of the following complexes:
   a) bis(ethylenediamine)dichlorochromium (III)
   b) potassium hexafluoronickelate(IV)
   c) $[Cu(NH_3)_4]SO_4$
   d) $Na_4[Fe(CN)_6]$

4. How many compounds have the formula $[CoCl_2(NH_3)_4]$? Sketch the compounds and label appropriately.

5. Experimentally how might you distinguish between the two compounds, $[Co(NH_4)_2(SO_4)_2]Cl$ and $[Co(NH_4)_2(Cl)_2]SO_4$? Use common laboratory reagents and explain.

Key Concepts for Free Response

1. In general the density increases as you proceed down the chart in the 4th, 5th, and 6th
   Periods. The atomic volume does not change much as the masses increase, so the densities
   increase. Moving across a period, the most dense elements are in the middle section, group
   8B because electrons are being added to orbitals with already have one electron.
   The melting points of the transition metals are highest in the middle of a series when the d
   subshell is about half-filled. In general as the atomic mass increases down the chart from the
   4th, to 5th to 6th period, the melting points increase.

2. a) pyrometallurgy
   b) hydrometallurgy
   c) bacteria

3. a) $[Cr(en)_2Cl_2]^+$
   b) $K_2[NiF_6]$
   c) tetraamminecopper(II) sulfate
   d) sodium hexacyanoferrate(II)

4. The octahedral complex has six ligands about the Co^{2+} which can be in a cis or a trans
   arrangement. Optical isomers would not be expected for this compound.

5. One compound, $[Co(NH_4)_2(SO_4)_2]Cl$, will have a chloride ion in solution, so the addition of
   $AgNO_3$ will cause the precipitation of AgCl. The other compound has the chloride ions as
   ligands and they are not free to react. No precipitate will form. Similarly, the sulfate group
   is free to react in a solution of $[Co(NH_4)_2(Cl)_2]SO_4$ so a precipitate will form with $BaCl_2$.

Section B: Multiple Choice

6. Which of the following groups of elements occur in nature in the free state?
   a) Na, K, Li   b) Fe, Co, Mn   c) Sc, Ti V
   d) Pb, Bi, Sn   e) Au, Pt, Ir

7. Which of the following transition metal compounds is used as a white paint pigment?
   a) CoO_2 b) $CuSO_4$ c) NiO_2 d) Cr_2O_3 e) TiO_2

8. In a blast furnace, iron oxides are reduced by
   a) $CaCO_3$ b) CO   c) CO_2 d) $CaSO_4$ e) CS_2

9. In a blast furnace, iron oxides are reduced by
   a) $CaCO_3$ b) CO   c) CO_2 d) $CaSO_4$ e) CS_2

10. Which of the following metals is most dense?
    a) Cr   b) Ta   c) W   d) Hg   e) Mo

11. Which element would have the highest melting point?
    a) Cd   b) Ru   c) Ta   d) Hg   e) Os

12. In a blast furnace the most effecting reducing agent for Fe_2O_3 is
    a) SiO_2 b) H_2 c) CO   d) CO_2 e) Al

13. Pig iron is
    a) a soft and brittle form of iron that contains impurities
    b) a hard and ductile form of iron that contains impurities
    c) a hard very pure form of iron
    d) a soft very pure form of iron
    e) a partially oxidized form of iron

14. Of the ions, Cr^{2+}, Zn^{2+} and Ni^{2+} , which is (are) diamagnetic?
    a) Zn^{2+} only   b) Cr^{2+}only   c) Ni^{2+}only
    d) Zn^{2+} and Cr^{2+} e) all three ions

15. Copper ores are enriched to increase the percentage of copper in the mixture by a process called
    a) roasting   b) flotation   c) pyrometallurgy
    d) hydrometallurgy   e) ganguation

16. What is the oxidation number of the metal ion in $[Co(NH_3)_5SO_4]Cl$
    a) +1   b) +2   c) +3   d) +5   e) +6

17. What is the oxidation number of the metal ion in $[Pt(NH_3)_2]Cl_2$
    a) 0   b) +2   c) +4   d) +6   e) -1

18. An example of a neutral bidentate ligand is
    a) ammonia   b) oxalate $(C_2O_2^{2-})$ c) acetate
    d) ethylenediamine   e) EDTA

19. The chelating agent EDTA, ethylenediaminetetraacetic acid, is classified as a
    a) a neutral bidentate ligand
    b) a neutral hexadentate ligand
    c) a tetradentate ligand with a 2- charge
    d) a tetradentate ligand with a 6- charge
    e) a hexadentate ligand with a 4- charge

20. The name of the coordination compound with the formula is $Fe(CO)_5$ is
    a) iron (V) carbonmonoxide
    b) iron(III) pentacarbonmonoxate
    c) iron (III) pentacarbonmonoxide
    d) pentacarbonyliron(0)
    e) pentacarbonyliron(II)

21. The name of the coordination compound with the formula $(NH_4)_2[CuCl_4]$
    a) ammonium tetrachlorocuprate(II)
    b) diammonium copper (II) tetrachloride
    c) ammonium copper (II) chloride
    d) diammonium tetrchlorocopper(II)
    e) copper (II) diamminetetrachloride

22. The name of the coordination compound with the formula $Na[FeCl_4]$ is

    a) sodium iron(III) tetrachloride

    b) sodium tetrachloroferide(III)

    c) sodium chloroferrate(IV)

    d) sodium tetrachloroferrate(III)

    e) sodium ferroyltetrachloride

23. The name of the coordination compound with the formula $Na[FeCl_4]$ is

    a) sodium ferroyltetrachloride

    b) sodium iron(III) tetrachloride

    c) sodium tetrachloroferide(III)

    d) sodium chloroferrate(IV)

    e) sodium tetrachloroferrate(III)

24. The name of the coordination compound with the formula $[Co(en)_3]Cl_3$ is

    a) (ethylenediamine)cobalt(III) chloride

    b) tris(ethylenediamine) cobalt(III) chloride

    c) ethylenediaminecobalt trichloride

    d) tris(ethylenediamine) cobalt trichloride

    e) (ethylenediamine) cobalt (IV) chloride

25. The formula for triamminedichloronitritocobalt (III) is

    a) $[Co(NH_3)ClNO_2]$ b) $[Co(NH_3)_3ClNO_2]$ c) $[Co(NH_3)_3Cl_2NO_2]$

    d) $[Co(NH_3)_3(Cl_2)_2NO_2]$ e) $[Co(NH_3Cl_2NO_2)_3]$

26. The formula for the hydroxopentaaquairon (III) ion is

    a) $[Fe(OH)(H_2O)_5]^{3+}$ b) $[Fe(OH)(H_2O)_5]^{2+}$ c) $[Fe(OH)_5]^{3+}$ (aq)

    d) $[(H_2O)_5Fe](OH)_3$ e) $[Fe·5 H_2O](OH)_3$

27. Which of the following can form geometric isomers?

    a) $[Co(NH_3)_6] Cl_3$ b) $[Co(NH_3)_5Cl]^{2+}$ c) $[Co(NH_3)_4Cl_2]^+$

    d) $[Co(NH_3)_5Cl]SO_4$ e) $[Co(Cl)_6]^{4-}$

28. Which of the following can form optical isomers?
   a) $CHCl_3$ b) CH_2Cl_3 c) $(CH_3)(CH_2CH_3)_2NH^+$
   d) $BrCH(CH_3)CO_2H$ e) $CH_2(CO_2H)_2$

29. How many unpaired electrons are present in the high spin complex $[Fe(H_2O)_6]^{2+}$
   a) 8   b) 6   c) 4   d) 2   e) 0

30. Which of the following has a d5 electron configuration?
   a) $Co(CN)_4^-$ b) $Mo(NH)_3^{3+}$ c) $Rh(Cl)_6^{4-}$
   d) $V(CN)_6^{4-}$ e) $Fe(CN)_6^{3-}$

Chapter 23 : Answers to Multiple Choice

6. e	16. c	26. b
7. e	17. b	27. c
8. b	18. d	28. d
9. b	19. d	29. c
10. c	20. a	30. e
11. e	21. d	
12. c	22. d	
13. a	23. e	
14. a	24. b	
15. b	25. c	

Chapter 24
Nuclear Chemistry

Section A: Free Response

NO   O K

1. List by symbol the three isotopes of hydrogen. Explain how a radioactive isotope differs
    from a non-radioactive isotope using hydrogen to illustrate the concept.

OK

2. Carbon-11 is an unstable isotope. Explain why it is unstable and suggest a type of
    radioactive decay that is expected.

NO

3. Discuss the reactions in which carbon-14 is formed from nitrogen in the upper atmosphere
    and eventually becomes part of the food chain on earth. Explain how it can be used to
    determine the age of some materials.

4. Complete and balance the following nuclear reactions.

OK

a) $^{113}_{47}Ag \rightarrow \beta + ?$
b) $^{16}_{8}O + ^{16}_{8}O \rightarrow \alpha + ?$
c) $^{235}_{92}U + ^{1}_{0}n \rightarrow ^{90}_{38}Sr + ? + 3 \, ^{1}_{0}n$

5. Sketch a plot which can be used to predict the stability of nucleons. Label the axes.

OK

Explain how nuclei not within the peninsula behave.

Key Concepts for Free Response

1. The three isotopes of hydrogen are ^1_1H, ^2_1H, and ^3_1H which are called protium (hydrogen), deuterium, and tritium. Each has one proton in its nucleus but deuterium also has one neutron and tritium has two neutrons. A radioactive isotope emits radiation which can be detected. Only tritium is radioactive. It my emit a beta particle to contain 2 protons and 1 neutron through the process

$$^3_1H \rightarrow ^3_2He + ^0_{-1}e$$

2. Carbon 11 has 6 protons and only 5 neutrons. It would be expected to emit a positron so that a proton is transformed into a neutron to produce boron-11

3. Cosmic rays break apart gases in the upper atmosphere which release neutrons. These can bombard nitrogen nuclei to form carbon 14 in very very small amounts by the equation:

$$^{14}_7N + ^1_0n \rightarrow ^{14}_6C + ^1_1H$$

The radioactive carbon eventually reacts with oxygen to form CO2 and is incorporated into plants and the food chain on earth. The amounts are extremely small, but measurable. When an organism is alive and constantly rebuilding molecules the amount remains constant but after it dies, the carbon-14 decreases and can be used to determine the age of very old but once living materials.

4. a) $^{113}_{47}Ag \rightarrow \beta + ^{113}_{46}Pd$

   b) $^{16}_8O + ^{16}_8O \rightarrow \alpha + ^{28}_{14}Si$

   c) $^{235}_{92}U + ^1_0n \rightarrow ^{90}_{38}Sr + ^{143}_{54}Xe + 3\ ^1_0n$

5.

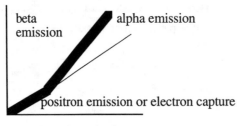

Number of
neutrons

beta emission   alpha emission

positron emission or electron capture

Number of protons

Section B: Multiple Choice

6. Which of the following symbols is used to represent gamma ray emissions?

    a) $_{1}^{-1}\gamma$ b) $_{0}^{0}\gamma$ c) $_{2}^{4}He$ d) $_{4}^{2}He$ e) $_{-1}^{+1}\gamma$ *OK*

7. Which of the following symbols is used to represent alpha particle emissions?

    a) $_{1}^{-1}\gamma$ b) $_{0}^{0}\gamma$ c) $_{2}^{4}He$ d) $_{4}^{2}He$ e) $_{-1}^{+1}\gamma$ *O/c*

8. Which of the following symbols represents positron emissions?

    a) $_{2}^{4}He$ b) $_{4}^{2}He$ c) $_{-1}^{0}e$ d) $_{+1}^{0}e$ e) $_{0}^{0}e$ *O/c*

9. Which type of radiation is effectively blocked by a piece of paper?   *NO*

    a) alpha particles   b) beta particles   c) electron capture

    d) positron emissions   e) gamma radiation

10. The atomic number of a nucleus which emits a beta particle will

    a) remain the same   b) increase by one unit   c) increase by two units   *OK*

    d) decrease by one unit   e) decrease by two units

11. The atomic number of a nucleus which emits a positron will

    a) remain the same   b) increase by one unit   c) increase by two units   *dK*

    d) decrease by one unit   e) decrease by two units

12. The mass number of a nucleus which emits a positron will

    a) remain the same   b) increase by one unit   c) increase by two units   *OK*

    d) decrease by one unit   e) decrease by two units

13. The atomic number of a nucleus which undergoes electron capture will   *OK*

    a) remain the same   b) increase by one unit   c) increase by two units

    d) decrease by one unit   e) decrease by two units

14. If a nucleus decays by successive α, α, β decay, the atomic number will
    a) increase by four units   b) increase by three units   c) increase by 1 unit
    d) decrease by eight units   e) decrease by three units

15. If a nucleus decays by successive α, β, β emissions, how would the atomic number and mass number change?
    a) The atomic number decreases by four; the mass number stays the same.
    b) The atomic number increases by two; the mass number decreases by two units.
    c) The atomic number stays the same; the mass number decreases by two units.
    d) The atomic number decreases by two; the mass number decreases by four units.
    e) The atomic number stays the same; the mass number decreases by four units.

16. If a nucleus undergoes successive α, α, β decay, the atomic number of the substance produced will
    a) increase by four units   b) increase by three units   c) increase by 1 unit
    d) decrease by eight units   e) decrease by three units

17. When uranium-235 undergoes successive α, β, β , α emissions, the nucleus produced is
    a) ^{84}Po   b) ^{230}Th   c) ^{227}U   d) ^{230}Ra   e) ^{228}Pa

18. When lead-210 undergoes successive β, β , α emissions, the nucleus produced is
    a) ^{206}Pb   b) ^{214}Rn   c) ^{210}Po   d) ^{204}Tl   e) ^{228}Pa

19. The nucleus b) $^{53}_{24}$Cr is produced by the beta decay of
    a) $^{53}_{25}$Mn   b) $^{54}_{24}$Cr   c) $^{52}_{24}$Cr   d) $^{53}_{23}$V   e) $^{54}_{23}$V

20. What are the labels for the x and y axes respectively for the peninsula of stability graph?

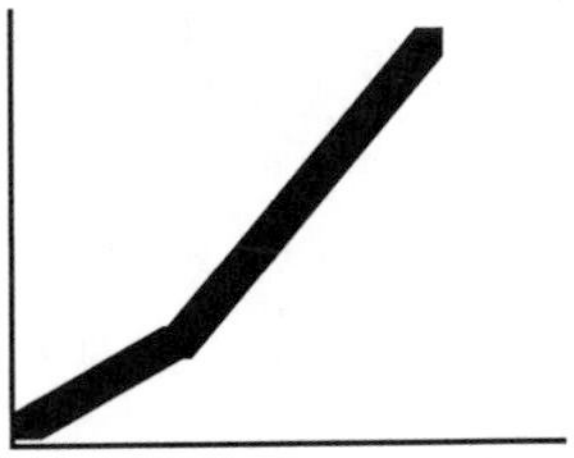

a) number of protons, number of neutrons
b) number of protons, number of electrons
c) number of neutrons, number of nucleons
d) number of protons, number of nucleons
e) number of neutrons, mass number

21. Elements above the peninsula of stability generally emit
a) alpha particles because they have too many neutrons to be stable
b) beta particles because they have too many neutrons to be stable
c) positrons because they have too many neutrons to be stable
d) alpha particles because they have too many protons to be stable
e) beta particles because they have too many protons to be stable

22. Elements below the peninsula of stability may
a) attain stability by beta emission because they have too few neutrons.
b) attain stability by beta emission because they have too many neutrons.
c) attain stability by positron emission because they have too many neutrons.
d) attain stability by positron emission because they have too few neutrons.
e) attain stability by electron capture because they have too many neutrons.

23. Complete the following fission reaction:

$$^{235}_{92}U + ^{1}_{0}n \rightarrow ^{139}_{53}I + \underline{\quad ? \quad} + 2\,^{1}_{0}n$$

a) $^{95}_{39}Y$ b) $^{96}_{39}Y$ c) $^{96}_{40}Zr$ d) $^{95}_{40}Zr$ e) $^{94}_{41}Nb$

24. Complete the following nuclear reaction:

$$^{238}_{92}U + ^{12}_{6}C \rightarrow \underline{\quad?\quad} + 6\,^{1}_{0}n$$

a) $^{249}_{99}Es$ b) $^{249}_{98}Cf$ c) $^{244}_{92}U$ d) $^{244}_{98}Cf$ e) $^{250}_{104}Rf$

25. Complete the following fission reaction:

$$^{235}_{92}U + ^{1}_{0}n \rightarrow ^{139}_{53}I + \underline{\quad?\quad} + 2\,^{1}_{0}n$$

a) $^{95}_{39}Y$ b) $^{96}_{39}Y$ c) $^{96}_{40}Zr$ d) $^{95}_{40}Zr$ e) $^{94}_{41}Nb$

26. What will complete the following equation?

$$^{9}_{4}Be + ^{4}_{2}He \rightarrow + \underline{\quad?\quad} + ^{1}_{0}n$$

a) $^{13}_{7}N$ b) $^{14}_{7}N$ c) $^{12}_{6}C$ d) $^{13}_{6}C$ e) $^{14}_{6}C$

27. What particles are produced in the following reaction?

$$^{238}_{92}U + ^{16}_{8}O \rightarrow ^{250}_{100}Fm + \underline{\quad?\quad}$$

a) 2 neutrons   b) 4 neutrons   c) 1 alpha particle

d) 2 alpha particles   e) 4 alpha particles

28. What is the half-life of an isotope if the decay constant is 5.5 /day?

a) 0.13 days   b) 7.9 days   c) 0.36 days

d) 2.8 days   e) 5.5 days

29. What is the half-life of an isotope if the decay constant is 3.2/year?

a) 0.58 year   b) 1.6 year   c) 0.21 year

d) 8.7 year   e) 6.4 year

30. The half-life of $^{14}_{6}C$ is 5730 years. If a tree dies and lies undisturbed for 18,400 years. What percentage of the $^{14}_{6}C$ remains?

a) 19.2%   b) 17.4%   c) 10.8%   d) 8.43%   e) 0.053%

31. The half-life of ^{90}Sr is 28 years. How long will it take for a sample of ^{90}Sr to be 85% decomposed?

a) 23 years   b) 77 years   c) 83 years   d) 94 years   e) 110 years

32. The half-life of ^{32}P is 14.3 days. How much of a 15.0 gram sample of ^{32}P will remain after 75 days?

    a) 5.5 g   b) 3.7 g   c) 2.63 g   d) 0.64 g   e) 0.40 g

33. The half-life of iodine-131 is 8.0 days. If you have 25.0 grams of iodine-131, how much will remain after 40 days?

    a) 20.0 g   b) 5.00 g   c) 3.12 g   d) 0.781 g   e) 0.0390 g

34. The actual mass of oxygen-16 is 15.9949 g/mol. Calculate the binding energy of oxygen-16 based on the given information. Mass of a proton is 1.00783 g/mol. Mass of a neutron is 1.00867 g/mol. $c = 3.00 \times 10^8$ x/s   1 J= 1 kg·m^2/x^2

    a) 2.379×10^{13}J/mol   b) 2.379×10^8J/mol   c) 1.234×10^{13}J/mol

    d) 1.234×10^8J/mol   e) 1.734×10^8J/mol

35. Which of the following nuclear equations represent a fusion reaction?

    Reaction 1: $2\ {}^{18}_{1}H_2O \rightarrow 2\ {}^{3}_{1}H_2 \ + \ {}^{16}_{8}O_2$ OK

    Reaction 2: ${}^{2}_{1}H \ + \ {}^{3}_{1}H \rightarrow {}^{4}_{2}He \ + \ {}^{1}_{0}n$

    Reaction 3: ${}^{235}_{92}U \ + \ {}^{1}_{0}n \rightarrow {}^{144}_{54}Xe \ + \ {}^{90}_{38}Sr \ + 2\ {}^{1}_{0}n$

    a) Reaction 1   b) Reaction 2   c) Reaction 3

    d) Reactions 1 and 2   e) Reactions 2 and 3

36. What role do cadmium rods play in nuclear reactors?

    a) provide an alternative nuclear fuel

    b) give off electrons for the initiation

    c) absorb neutrons to control the rate of fission.

    d) fuse with ruthenium to initiate the reaction

    e) serve as an inert matrix for the reaction

37. What is a plasma?

    a) the transient product of a fission reaction

    b) unbound nuclei and electrons which may undergo fusion reactions easily.

    c) a liquid form of a radioactive element

    d) an unstable nucleus which can easily undergo fission

    e) a group of positrons

38. Which elements are in the greatest abundance in the sun?

    a) hydrogen and oxygen   b) hydrogen and helium   c) helium and tritium

    d) helium and deuterium   e) deuterium and tritium

39. Artificial (synthetic) radioactive isotopes of many elements used in medical treatments

    a) have extremely long half-lives

    b) are virtually non-existent

    c) can be made by bombardment of the element with neutrons.

    d) can be made in the depth of the earth only

    e) are produced within living tissues

40. In what year was the first nuclear bomb used as a weapon?

    a) 1925   b) 1945   c) 1955   d) 1965   e) 1975

41. Of these chemists, which one did not play an active role in our current understanding of nuclear chemistry?

    a) Bohr   b) Meitner   c) Einstein   d) Boyle   e) Fermi

42. The following persons all completed a college degree as a chemistry major. Which one continued as a professional chemist and was awarded a Nobel Prize in 1951?

    a) Janet Reno   b) Knute Rockne   c) Glenn Seaborg

    d) Margaret Thatcher   e) Paul Treichel

Chapter 24 : Answers to Multiple Choice

6. b	16. e	26. d
7. c	17. b	27. b
8. d	18. a	28. a
9. a	19. d	29. c
10. b	20. d	30. c
11. d	21. b	31. b
12. a	22. d	32. e
13. d	23. a	33. d
14. e	24. d	34. c
15. e	25. a	35. b

36. c

37. b

38. b

39. c

40. b

41. d

42. c